Climate Change

A Beginner's Guide

363.73874 B789c 2010
Boyd, Emily
Climate change : a
beginner's guide

WITHDRAWN

CUYAHOGA COMMUNITY COLLEGE
EASTERN CAMPUS LIBRARY

ONEWORLD BEGINNER'S GUIDES combine an original, inventive, and engaging approach with expert analysis on subjects ranging from art and history to religion and politics, and everything in between. Innovative and affordable, books in the series are perfect for anyone curious about the way the world works and the big ideas of our time.

anarchism
aquinas
artificial intelligence
the beat generation
biodiversity
bioterror & biowarfare
the brain
the buddha
censorship
christianity
civil liberties
classical music
climate change
cloning
cold war
conservation
crimes against humanity
criminal psychology
critical thinking
daoism
democracy
dyslexia
energy
engineering
the enlightenment
evolution
evolutionary psychology
existentialism
fair trade
feminism
forensic science
french revolution
history of science
humanism
islamic philosophy
journalism
lacan
life in the universe
literary theory
machiavelli
mafia & organized crime
magic
marx
medieval philosophy
middle east
NATO
oil
the palestine–israeli conflict
philosophy of mind
philosophy of religion
philosophy of science
postmodernism
psychology
quantum physics
the qur'an
racism
renaissance art
the small arms trade
sufism

Climate Change

A Beginner's Guide

Emily Boyd and Emma L. Tompkins

ONEWORLD
OXFORD

CUYAHOGA COMMUNITY COLLEGE
EASTERN CAMPUS LIBRARY

A Oneworld Paperback Original

Published by Oneworld Publications 2009

Copyright © Emily Boyd and Emma L. Tompkins 2010

The right of Emily Boyd and Emma Tompkins to
be identified as the Authors of this work has been
asserted by them in accordance with the
Copyright, Designs and Patents Act 1988

All rights reserved
Copyright under Berne Convention
A CIP record for this title is available
from the British Library

ISBN 978–1–85168–660–5

Typeset by Jayvee, Trivandrum, India
Cover design by Simon McFadden
Printed and bound in Great Britain by

Oneworld Publications
185 Banbury Road
Oxford OX2 7AR
England
www.oneworld-publications.com

Learn more about Oneworld. Join our mailing list to
find out about our latest titles and special offers at:

www.oneworld-publications.com

For our families and godchildren

Contents

Preface

The key messages of this book are that climate change is happening and action needs to be taken. For thirty years, scientists, businesses, politicians and non-governmental organisations have debated first the reality, and now the causes, of climate change. The stakes are high.

In March 2008, a group of leading climate scientists wrote in the journal *Science* that the European Union's decision to try to keep the rise in global temperatures to below 2°C was 'a recipe for global disaster' and that the rise needed to be far lower than this. To keep warming below this figure, scientists argue that we must ensure the concentration of carbon dioxide in the atmosphere does not go above 450 parts per million (ppm). If levels of carbon dioxide – one of the most significant 'greenhouse gases' – are allowed to rise above this level, the Earth's climate could go out of balance. However, carbon dioxide concentrations have already increased from 280ppm in the middle of the nineteenth century to 385ppm in 2008.

The Earth's climate may already have passed a critical point of no return. In 2005, at a conference on climate change organised by the UK Meteorological Office, scientists warned politicians that although we should aim to keep warming down to 2°C, we need to prepare society for a 4°C rise. A four-degree rise will result in a vastly different world: dangerous water shortages, frequent storms, drought, and completely altered human and physical geographies. There is no simple solution. We need to think creatively about how we can live with the worst consequences of climate change and ask ourselves difficult questions. How can societies adapt? Who will adapt? What will we need to adapt to?

There are already lucrative business prospects in global trading in greenhouse gases: climate change will force us to think of ways to transform societies and confront the limits of adaptation. For many, the impacts of climate change will be severe; for others, it will provide opportunities. Climate change will not only exacerbate climate and weather-related hazards but also the social maladies caused by the age-old problems of corruption, poverty, injustice and inequality. Politicians and public servants face the challenge of making difficult decisions without knowing which scenario is likely to unfold. The limited financial and human resources of local and national governments may hamper their ability to come up with innovative and cost-effective climate solutions for their constituents. Governments need to act; to pursue stringent policies to stabilise greenhouse gas levels and to create incentives for changes in behaviour. But they can only do so if people, businesses and the media give them a strong enough signal.

The science and politics of climate change are complex; but humans live and cope with complications and uncertainty every day. The only way to tackle the problem is to be guided by confident caution, not paralysed by fear or uncertainty.

Acknowledgements

We are both lucky to have received guidance on climate change from some of the best thinkers in the world, including Professor Mike Hulme at the University of East Anglia and Professors Katrina Brown and Neil Adger at the Tyndall Centre for Climate Change, and direct access to experts on sustainability science such as Professors Tim O'Riordan and Carl Folke. Further inspiration has come from Professor Diana Liverman – who has, since the early 1980s, been a tower of strength in championing climate change and development. We both owe so much to these people for their direction and inspiration. Other scholars from a range of disciplines that have inspired us include Professor Daniel Kahneman, Professor Susan Owens, Professor Ronald Mitchell, Professor Elinor Ostrom and Professor Andy Gouldson.

We feel privileged to have had the opportunity to spend time as Fellows at the Oxford University Centre for the Environment (OUCE), assisted by the generous funds from the James Martin 21st Century School and the Leverhulme Trust. In this context we extend our gratitude to our fellow James Martin colleagues with whom we have worked closely and shared special times. In particular, we want to acknowledge Maxwell Boykoff for enlightening us on the cultural interpretations of climate change and the media. Great thanks go to David Frame, our resident 'paper clip' and climate scientist *extraordinaire*, Samuel Randalls, the heretic who taught us all we need to know about the world of constructs, Maria Carmen Lemos who raises the bar on the notion of political theory, Nathan Hultman, for his intellectual capacity, rigour and enthusiasm and Timmons Roberts, who highlighted for us the importance of justice and equity in climate change and also managed to produce several books during his brief sojourn in Oxford.

The Environmental Change Institute in the OUCE has been our home during the time of writing this book. The Institute provided an extremely supportive environment throughout. Special thanks go to Deborah Strickland for drawing illustrations for the book, Joshua Knowles for his illustration of the politics of climate change (Chapter 5) and Maria Mansfield for her constructive comments and contributions. We are grateful to Lucy Hayes who volunteered assistance and produced example boxes that are included in the book. Other institutions, which deserve a special mention, are the Sustainability Research Institute at Leeds University and the Stockholm Resilience Centre at Stockholm University – both institutions have provided us with a wealth of intellectual support. A particular note of thanks to our colleagues and friends: Polly Ericksen, Heike Schroeder, Robert Hope, Mark New, Yadvinder Malhi, Simon Batterbury, Scott Prudham, Jimin Zhao, Dave Stainforth and Arthur Mol – just a few of the people with whom we have been fortunate to interact during our time at Oxford.

Special thanks go to Richard Walters for his constant support and constructive criticism and for clarifying the rationale of models. Bo Kjellén, an inspirational figure for many of us working on climate change, kindly provided edits on the content of the book, in particular on the politics of climate change. We are grateful to John Boyd for his help with the grammatical and linguistic aspects of the book. Finally, special thanks are extended to Marsha Filion who has been an extremely supportive publisher and colleague.

Introduction

> 'Now is not the time for half measures. It is the time for a revolution.'
>
> (President Jacques Chirac, cited in *Nature,* 8 February 2007)

Many books on climate change describe only the science. This book focuses on the impacts of climate change on societies, their direct and indirect consequences and the opportunities to address those consequences.

We want to present the facts, politics, people and problems associated with climate change in a straightforward, non-sensational way and communicate clearly why we need to do something *now.* An April 2009 survey of 261 international climate scientists, carried out for the *Guardian*, showed that nine out of ten scientists, who are all too aware of the risks from climate change, believe that governments will fail to keep global warming rises below the potentially dangerous increase of 2°C. Yet people in many countries still do not acknowledge the urgency of the problem: a gulf exists between scientists' and public understanding of climate change. In Australia, a 2008 poll by the Lowry Institute showed little public support for urgent action on climate change. While 60% of those polled agreed that 'we should begin taking steps now even if this involves significant costs', compared to the 2006 survey, there was a declining willingness to pay those costs. In the UK, an Ipsos MORI poll of over one thousand adults, also taken in 2008, showed that the majority of the British public are still not convinced that climate change is the result of human activity and believe that scientists are exaggerating the issue. Interestingly, in the US, an Ellison Research poll found that 89% of Evangelical Christians believed that the US should seek to 'curb its global warming pollution, regardless of what other nations do'. Yet this segment of society does not appear to be representative of wider American public attitudes to climate change: a March 2009 Gallup survey of 1,012 people showed an increasing number think that the risks of climate change have been exaggerated. The challenge for academics is to find ways of getting the message across to those groups in society that are currently not able to access the science.

We believe climate change is a problem caused by many people's actions over a long time. Sorting out the problem will require collective solutions that span households, communities and nations; co-ordinated actions across networks, civil society, institutions and scientific disciplines. Huge social transformations will be required; institutional barriers will have to be removed and scientific silos will have to be reconfigured. The evidence of previous successful social transformations suggests that such social, cultural, scientific and institutional change is possible – and beneficial in the long term. People have reshaped their societies and ways of living after discovering fire, after inventing wheels, in the transition from sailing ships to steam ships, from using horse-drawn carriages to motor cars, from muscle-powered to electrically-powered machinery. Each of these transitions required significant shifts in attitudes and in institutions. There have always been those who disagreed with the need for change but change has been managed and society has continued to grow and to develop. These historical examples give us hope that we will be able to manage the next big transition, the shift away from carbon-intensive industries that will allow us to stabilise greenhouse gas levels.

Stabilising carbon dioxide levels at a point that avoids dangerous climate change (whether this is 350, 450 or 500ppm) will require a rapid curbing of emissions of this greenhouse gas. We need to change contemporary large-scale industrial methods and energy consumption and address the complex links between land use in the tropics and consumer choice in the developed world. Effective solutions will mean adapting old technology and transforming energy-intensive industries; developing new technology – such as experimenting with carbon capture and storage (in which carbon dioxide is extracted from the atmosphere and stored underground or under the sea); and finding substitutes for fossil fuels.

Science will play a prominent role in describing climate change and predicting its impacts, and in developing and evaluating different adaptations. However, we foresee that much of the impetus for real action will come from business: new initiatives such as the creation of carbon markets and carbon offsetting. Carbon offsetting is a description of various schemes in which users of energy-intensive goods and services, such as flying, can pay a fee to compensate for the CO_2 emissions created. The fees are used to support projects that reduce the emission of greenhouse gases. Standards must be created so that voluntary carbon offsetting schemes become more effective and guarantee that offsetting projects deliver environmental, social and developmental benefits. Governments must ensure that public institutions, including universities, have sufficient resources to support continued research and the exploration of new ground-breaking ideas worthy of Nobel Prizes.

The biggest risk to society is that if the world suffers a major climatic impact (such as a major flood, drought or storm) the political desire to show decisive leadership will result in inappropriate and misdirected responses that waste resources. Investment in public institutions is needed, to give a clear signal to the private sector. Without clear government direction, private companies will flounder in a sea of confusion. Will oil, coal and gas producers and exporters continue to be able to extract and sell from current reserves of fossil fuels? Will tariffs be added? Will oil, coal and gas companies simply trade as much as possible on existing markets before they become unsaleable?

There are numerous inspirational books on climate change. Many provide an account of what might happen in the future and the potential impacts of different scenarios, but few provide much guidance on mitigating the effects of climate change or how to adapt to what we might expect. We hope that this book, with its focus on what to do and how to prepare, goes some way towards filling the gap. Our chapters are structured around the key questions of climate change, explaining what we need to know about the science. What is the phenomenon of climate change and why is it such an urgent problem? Why does so much confusion surround climate change, its science and its expected impacts? Why is it so hard for politicians to take climate change seriously? How do we move from political game-playing to responsible management? What will happen if we don't address the problem? Who will be the climate change winners and losers? How can we ensure that everyone on the planet has a chance of being able to cope with its consequences?

It's easy to be pessimistic about climate change or to think that climate scientists make exaggerated claims. This book offers a frank and thorough explanation of the facts, the uncertainties, the politics, the options and the risks, to allow us all to make up our own minds.

1 The climate is changing

> 'Today, the time for doubt has passed. The IPCC has unequivocally affirmed the warming of our climate system and linked it directly to human activity.'
>
> (Ban Ki-Moon, United Nations Secretary General, September 2007)

There is no doubt: the world is getting warmer.

This warming is largely caused by greenhouse gas emissions associated with human activity. The more than two and a half thousand scientists who comprise the international Intergovernmental Panel on Climate Change (IPCC) have concluded that climate change is a truly global phenomenon that requires global action – a challenge that mankind has never before had to confront.

Humans have been adapting to a changing Earth for millennia, but the revolution we will have to face from climate change is unprecedented. Human creativity and ingenuity has improved our lives and enhanced our development prospects through our industries, making greenhouse gases (water vapour, carbon dioxide, methane, nitrous oxide and hydrofluorocarbons) along the way. Now, we know that these gases are stored in the atmosphere, where they act like a blanket, preventing the sun's reflected heat from leaving the atmosphere. Not all gases remain in the atmosphere for the same time: nitrous oxide has an effective lifetime of about 100–150 years, carbon dioxide about 100 years and methane twelve years. We will have to live with increasing concentrations of these gases in the atmosphere – and hence increased warming of our planet – for many generations to come.

The Industrial Revolution, which started in Britain in the late eighteenth and early nineteenth century, was a turning point for our planet. During this period, national economies shifted from agricultural, driven by manual labour, to industry, driven by new technologies powered first by water and later by steam. Technology gave us more and more ways to use energy to make life easier, faster and richer: steam-powered ships, railways and factories; internal combustion engines; and the generation of electric power. Only now, as we see increasing concentrations of greenhouse gases in the atmosphere, are we realising the true cost of these fossil-fuel-powered technological innovations.

Since the mid-nineteenth century, the atmospheric concentration of carbon dioxide – one of the most important greenhouse gases – has increased from 280ppm to 385ppm and methane concentrations from 715 parts per billion (ppb) to 1774ppb. Geological evidence suggests that the Earth's temperature has been significantly higher, and carbon dioxide concentrations possibly as great as 450ppm, but these conditions occurred before the evolution of humans. We now expect the concentration of carbon dioxide in the atmosphere to rise above 450ppm before the end of the twenty-first century; we simply do not know what the Earth will be like when that happens.

The increase in atmospheric concentrations of greenhouse gases has been linked to their production by humans. In their Fourth Assessment Report, published in 2007, the IPCC estimated that between 1970 and 2004, greenhouse gas emissions increased by 70%. The largest increase, of 145%, was from the activities of the energy supply industry. Emissions from transport increased by 120% and from general industry by 65%. Other activities which are large emitters of greenhouse gases (see

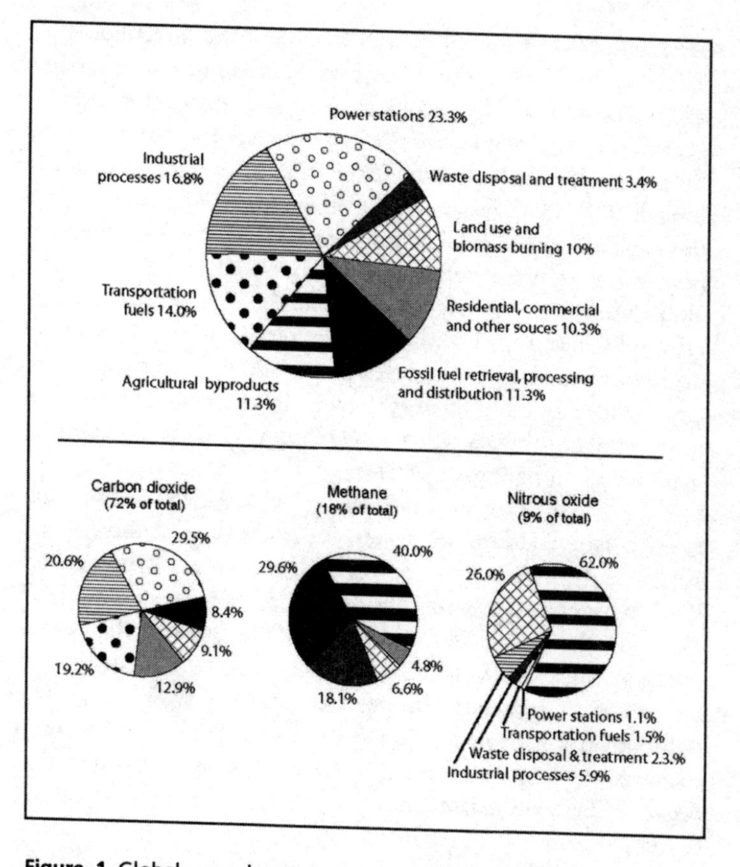

Figure 1 Global greenhouse gas emissions in 2000 (source: adapted from Emmanuelle Bournay, UNEP/GRID-Arendal)

Figure 1) include clearing land, changing land use, agriculture and building.

Projections of the future levels of greenhouse gas emissions have changed little over the past few years. This is not because of a lack of new information, but rather because present and past assessments match well, which suggests that we are very close to identifying the impact of continued increases in greenhouse gas levels. In 2001, when the IPCC produced its third major review of climate science and projections of future impacts, average global surface temperatures were expected to increase by between 1.5°C and 4.5°C by 2100. In 2007, using new data and models, the IPCC reassessed the situation. They now calculate the temperature will rise by between 1.1°C and 6.4°C, with the best estimates lying between 1.8°C and 4°C. It is clear that global warming is not stopping. While this will almost certainly cause significant problems for our planet over the coming decades, we must not forget that it may also bring opportunities: farmers may be able to grow new, higher-value crops, or be more productive with current ones, and new tourist ventures will emerge in new parts of the world.

As we will show, climate change is already causing rising average temperatures, sea level rises, increasing acidity of the oceans and more intense storms and sea surges. Everyone on Earth must consider what we need to do. 'We' are the rich and the poor; the people of the developed and those of the developing world. Politicians and scientists must keep the question of how to address climate change at the front of their minds, yet without creating panic. Climate change is a problem of risk: societies are vulnerable to the problems it causes, yet may also prove resilient and adaptable.

Climate change – a real and present danger?

People ask us: is climate change a natural phenomenon that has happened often on Earth? What has climate change got to do with freaky weather and what is the weather going to do next? How is climate change different from weather variation? Why is the science of climate change different now from how it was in the past? What are the risks of sudden climate change plunging us into a new Ice Age? Could there really be a 6°C increase in the Earth's temperature?

What are the barriers to our understanding the truth about climate change and why is there so much debate? Most people rely on the news media for information, yet the media give us many conflicting messages about climate change. The opinions delivered sometimes depend on the political interests of the proprietors, journalists and editors. Few journalists have any training in atmospheric chemistry and their lack of understanding, and desire to find attention-grabbing stories, combine to produce articles that too often hide the reality of the situation, either denying it is a problem or magnifying it into too large a problem.

Since the 1800s, scientists have known that there is a direct relationship between the concentration of greenhouse gases in the atmosphere and the average Earth temperature. The more carbon dioxide that is pumped into the atmosphere, the more the Earth's temperature goes up. This is simple physics, and is supported by the evidence of hundreds of climate models. It is indisputable. Yet, despite this long awareness of the evidence, until the mid-1980s the problems and challenges of climate change were largely unknown to people outside the research community. Contrast this with the early twenty-first century; now, even businesses compete to show their concern through public statements and research. In 2003, the huge insurance company Swiss Re expressed its concerns about the costs of climate change; in 2007, the bank HSBC funded a survey of 9,000 people on nine continents to assess their concerns about climate change; in the same year the financial institutions UBS and Citigroup published documents describing how climate change is shifting their exposure to risk.

These are just some of the many corporate initiatives trying to identify the impacts of climate change on business. But government and business concerns about climate change are not necessarily shared by the public. Popular fiction has even been written about the 'hoax' of climate change (such as Michael Crichton's *State of Fear*, published in 2004). Such 'climate change-denying' novels are written to sell, with themes of conspiracy and suggestions of hidden motives. However, *no* reputable scientists deny that our climate is changing. Climate change-deniers have been proved wrong and now look more like those who clung on to a belief that the Earth was flat long after it was proved to be spherical.

A handful of scientists remain 'climate change-sceptics' but they are a different breed to 'climate change-deniers'. They agree that the climate is changing but remain either mistrustful of the quality of climate models – suggesting that future predictions are too extreme – or argue that there might be other causes of change (such as sunspots) for which current climate models do not adequately account. Climate change-sceptics have an important role to play in the quest to understand climate change; their scepticism pushes other scientists into improving the quality of their climate models and identifying the ratio of human activity and other factors in contributing to climate change.

How did we get here and what do we know about the future?

Climate change exerts its effects on our lives in complex ways. Everything is connected. Producing greenhouse gases affects the global climate, which affects the weather and the way the oceans function (see Figure 2). The changing weather – storms, floods and other extremes – and ocean behaviour affect our capacity to grow food, find clean water or travel. Our response to these changes affects our emissions of greenhouse gases.

In 2007, the United Nations released three important reports, written by the IPCC, which made it clear that climate change is

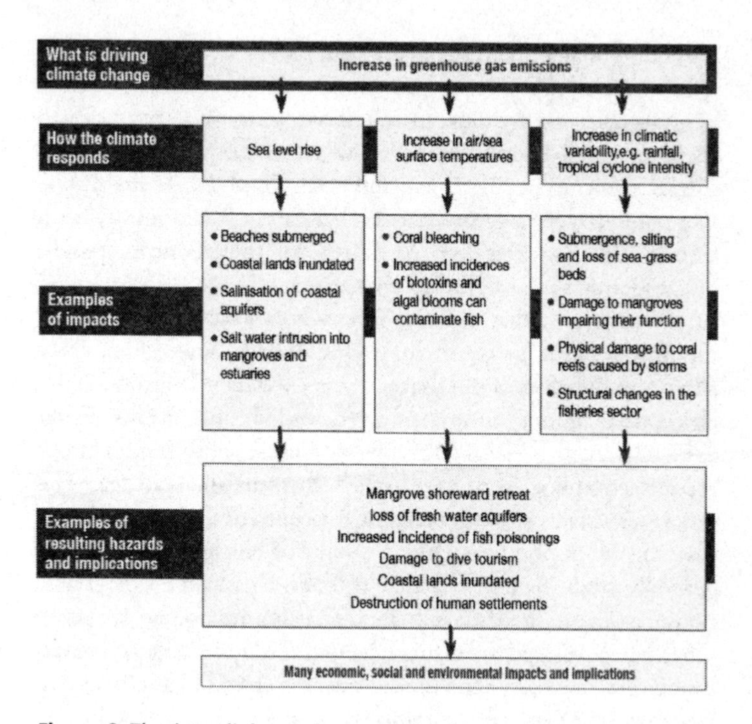

Figure 2 The inter-linkages between the human emissions of greenhouse gases and the changing coast (Source: Tompkins et al., 2005)

now 'unequivocal'. The IPCC was formed in 1988, under the auspices of the United Nations Environment Programme and the World Meteorological Organisation, to provide the best possible assessments of climate change. Scientists from the IPCC have measured, analysed and modelled many aspects of climate, including temperature, precipitation, storminess, extreme weather and weather hazards. Four sets of reports have now been produced (in 1990, 1995, 2001 and 2007). Each includes reports on the science of climate change, the impacts of climate change and proposed solutions. Each report is reviewed and revised by many scientists, many times, over its five-year preparation period. The reports' conclusions summarise the findings of more than two thousand scientists, whose contributions to these assessments are voluntary, and so are not the view of any single scientist. In 2007, the scientists involved in the IPCC were collectively awarded the Nobel Peace Prize (jointly with former US Vice-President Al Gore), for their contribution to our understanding of climate change.

According to the scientists who worked on the 2007 IPCC report, the Earth has warmed by about 0.75°C since 1860. They also point out that, of the thirteen years between 1995 and 2007, eleven are among the warmest recorded since 1850. Their evidence comes from dozens of high-quality temperature records compiled from data collected from land and sea. The report also states that the current concentrations of greenhouse gases in the atmosphere, and the rates of change, are unprecedentedly high. Scientists are able to find past levels of greenhouse gases in the atmosphere by analysing 'ice cores', long tubes of ice drilled from the ice of the Arctic and Antarctic. Using these ice cores, and the sediment extracted from them, we are able to assess concentrations of greenhouse gases in the atmosphere over the past million years. It is undeniable that concentrations of carbon dioxide, methane and nitrous oxide have increased significantly since the mid-nineteenth century, broadly since the Industrial Revolution in the developed world.

Information gathered from ice cores and sediments, in conjunction with techniques such as radiometric dating, ocean carbon levels, and other palaeo-climatic data, enables scientists to map the changes in global temperature over time and tell us how humans are affecting the Earth's climate. We know that the climate has been remarkably stable since the beginnings of human civilisation, about 6,000 years ago. And we also know that never before has the concentration of greenhouse gases in the atmosphere increased as quickly as it has done since the

1800s. No modern humans have ever witnessed such rapid change in the climate system as we are seeing now. This makes it extremely challenging to understand what is happening, particularly since we cannot see the changes at first-hand. It is also difficult to think conceptually or envisage how we can adapt and change quickly enough.

Scientists now have higher confidence not only in the projected patterns of warming but also in other features, including changes in wind patterns, precipitation, sea ice and extreme weather. Globally, water is a crucial issue, although the effects vary geographically. Scientists have observed an increasing frequency of heavy bursts of rain over most areas of the world. In 2005, in Mumbai, India, 94 centimetres of rain fell in one day; unprecedented in recorded history. Between 1900 and 2005, precipitation increased significantly in eastern parts of North and South America, northern Europe and northern and central Asia but declined in the Sahel, the Mediterranean, southern Africa and parts of southern Asia. Globally, the area affected by drought has increased since the 1970s. In Africa, by 2020, between 75 and 250 million people may suffer water shortages. In some areas, agricultural yields may be reduced by 50% if farmers do not shift to drought-resistant crops. In Asia, scientists estimate that by the 2050s, freshwater will be much less available and coastal areas will be at greater risk from sea inundation. Water shortages are also anticipated on small islands. Between 1961 and 1993, sea levels rose more quickly than ever before. In 1961, sea-level rises were about 1.8mm a year; by 1993, annual average sea level rise was 3.1mm. These rises are mostly due to thermal expansion of the oceans (as liquids become warmer, they expand). As the atmosphere has become warmer, this has warmed the surface layers of the oceans and they have increased in volume. The melting of glaciers, ice caps and polar ice sheets also affects sea levels. Scientists estimate that by the end of the twenty-first century, sea levels could rise by between 18 and

CLIMATE OR WEATHER?

We all know there is a difference between *climate* and *weather*. Weather is the day-to-day fluctuation that determines whether we carry an umbrella or wear a sunhat on our way to work. Irregularities, like an unusually hot day in winter or snow in summer, do not imply dramatic climate change – they are simply anomalies. We experience different kinds of daily weather in the different climate zones around the world. Temperate climate zones have four seasons (spring, summer, autumn and winter), each with their expected heat, rain and storms; tropical climate zones have wet and dry seasons, with very different average daily tempera-

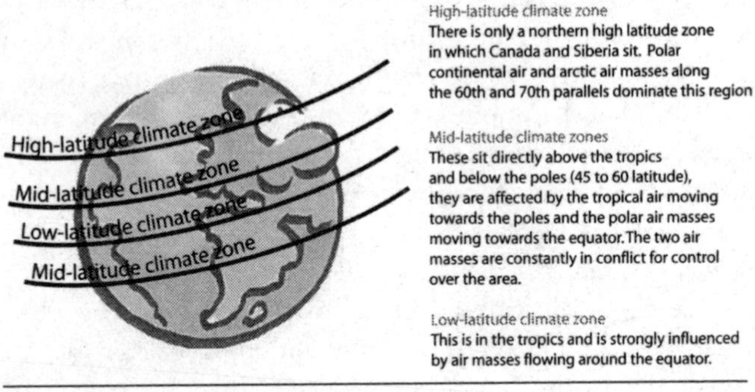

High-latitude climate zone
Mid-latitude climate zone
Low-latitude climate zone
Mid-latitude climate zone

High-latitude climate zone
There is only a northern high latitude zone in which Canada and Siberia sit. Polar continental air and arctic air masses along the 60th and 70th parallels dominate this region

Mid-latitude climate zones
These sit directly above the tropics and below the poles (45 to 60 latitude), they are affected by the tropical air moving towards the poles and the polar air masses moving towards the equator. The two air masses are constantly in conflict for control over the area.

Low-latitude climate zone
This is in the tropics and is strongly influenced by air masses flowing around the equator.

Figure 3 Climate zones around the Earth

tures. *Climate change* refers to significant shifts in weather over time: a wet season becoming consistently wetter, a dry season lengthening or the monsoon season becoming less predictable.

The Köppen Climate Classification System is a standard system used to describe the Earth's climates. In 1900, Wladimir Köppen classified the world's climatic regions, broadly in line with the global classification of vegetation and soils. In this system, which is based on temperature and precipitation, there are five main climate types: *Moist Tropical Climates* (high year-round temperatures and precipitation); *Dry Climates* (very little precipitation and

CLIMATE OR WEATHER? (*cont.*)

significant daily variation in temperatures); *Humid Middle Latitude Climates* (warm, dry summers and cool, wet winters); *Continental Climates* (low precipitation and significant seasonal variation in temperature); and *Cold Climates* (permanent ice and tundra and fewer months of the year above freezing than below). These five climate types can be found in the three planetary climate zones, delineated by the air masses that affect them.

59cm. The partial loss of ice sheets on polar land could bring an additional 20 to 60cm rise. Such rises will bring about major changes to coastlines and cause inundation of low-lying areas. River deltas and low-lying islands, presently inhabited by millions of people, will be very badly affected. Small island states, such as the Marshall Islands, are expected to experience increased inundation, storm surge erosion and other coastal hazards, which will threaten their essential infrastructure. Warmer oceans could also change the ocean currents (such as the Meridional Overturning Circulation) that carry warm water into far northern latitudes and return cold water southward. Large and persistent changes in these currents will affect marine ecosystem productivity, fisheries, ocean carbon dioxide uptake and land vegetation.

The effects of climate change experienced around the world will, to a large extent, be determined by the choices we are making now. We can make the problem far worse by pursuing lifestyles that require high levels of fossil fuels or we can lessen the impacts by adopting low-carbon lifestyles, saving energy and reducing our emissions. Our future impacts can also be reduced if we start planning today. We do not *have* to experience the terrible losses of wild storms or horrific floods. Yet the complex and subtle dance of leader and follower between the private sector and the state is, in many ways, slowing down these vital preparations. The private sector is waiting for leadership and guidance from the state, yet in most countries the state is reluctant to impose unpopular controls.

What is at stake?

Climate change is frequently framed as a problem of risk and vulnerability, rather than of impacts and responses. Such use of language contributes to the confusion about what is really happening. Understanding risk and vulnerability will significantly enhance our ability to prepare for climate change. They can perhaps be most easily understood by considering what they mean in specific regions and countries.

China has a large number of very poor people living in rural areas with limited natural resources, who are therefore sensitive to drought. Some parts of China are highly prone to drought simply because of their geography and topography; they just never receive much rain. The picture is similar throughout Southern Asia, making it particularly at risk to climate change. Vulnerability extends beyond geography. The after-effects of colonialism, such as democratisation, land-partitioning and the imposition of unnatural boundaries, have left densely-populated marginal countries (for example, Bangladesh, Pakistan and India) vulnerable to extreme climate-related stress. In all these countries, the melting of mountain glaciers is radically altering river flows and sea-level rises are drowning river deltas.

Neither is the developed world invulnerable to climate change, yet houses are still built in risky areas. The wealthy choose to live in beautiful yet fragile settings (such as coastal Florida), where their (often second) homes are at risk from winter storms, hurricanes and sea surges. On the other hand, the poor can be forced by circumstances into undesirable, but therefore cheap neighbourhoods (such as parts of New Orleans), prone to flooding, subsidence, storms and other hazards. Climate change is expected to increase the severity of weather-related hazards in years to come. As the world saw in the after-effects of Hurricane Katrina in 2005, lacking resources, the result for the poor was often catastrophic individual loss and homelessness, while the wealthy could afford to re-build or re-locate.

Holding a mirror to the past will not necessarily show us how climate change will affect us in years to come. Yet governments and businesses frequently make decisions about the future based on levels of risk inferred from past events. If the world is likely to be the same in the future as it was in the past, this can provide a good approximation of what may happen. But such models do not work as well when considering the effects of climate change. Planning for coping with the still-unpredictable effects of climate change means creating plans that can cope with uncertainty. In practical terms, this means considering a range of scenarios and leaving some flexibility to cope with surprises. More intense rainfall might lead to worse flooding, which might mean taking out more extensive flood protection insurance. Reinsurance companies like Swiss Re and Munich Re are already thinking about such adaptive planning.

Some are likely to lose from climate change but others will win. Societies can show resilience in the face of shock and surprise: they bounce back, re-organise, develop and adjust, without significant impairment of their function. Although it is hard to measure resilience, it can be seen in the ways that communities develop after a crisis. Societies can increase their resilience by developing mechanisms and institutions that enable them to remember past events, put early-warning systems in place, create buffers against the impacts of change and learn how to better prepare for future events. Such institutions vary in size, scale and reach. In the United States, the Federal Emergency Management Agency (FEMA) is charged with coping with disasters from preparation to recovery. FEMA's aim is to be ready to react to emergencies, to protect people's lives and property, and to provide information to support preparations for all kinds of emergencies, at all levels, throughout US society. The UK has no similar central emergency management body. Instead, specific agencies manage specific risks: for example, the National Health Service maintains plans for heat-wave emergency management and for coping with risks such as flu pandemics. Other government agencies are responsible for other high-profile risks, such as floods or animal disease epidemics. This structure has had some successes, such as the speedy and effective mobilisation of people and information in response to the foot and mouth disease outbreak in Surrey in 2007.

Regaining momentum after a crisis is fundamental to success. Lack of momentum can lead to disasters leaving behind long-term effects of despondency and social breakdown. FEMA's reaction (or lack of it) to the effects of Hurricane Katrina in 2005 illustrates this kind of unpreparedness. Multiple, complex and interacting forces meant FEMA's systems did not work effectively and failed to provide buffers against the effects of the disaster on the people of New Orleans. The city is undergoing an unnecessarily slow recovery from the devastation; four years on, it is still unclear how many people died and how many people remain homeless.

The choices we make

The lives we live in the twenty-first century are very different from the lives of our great-grandparents. They lived before television, refrigerators, mobile phones, space travel. Before full emancipation, the end of colonialism and the creation of new countries, the end of Soviet Communism and the re-appearance of many of the old Baltic States. Before two world wars and the nuclear age. Could our great-grandparents have predicted these changes? Are we likely to be any better? Any attempts to predict the future based on current information are likely to be erroneous. One thing we can be sure about is that people will adapt, whether to a changing climate or a changing society: humans are ingenious as well as fallible. We have little idea how people will cope with climate change but we can look back and consider what changes have happened and what adjustments have been made as a result of the small – 0.75°C – global temperature rise that has already taken place. Companies have started to include weather and climate factors in decision-making, an international trade in carbon emissions has started up, the UK has a government department for energy and climate change and climate change is written into international conventions.

We argue that there needs to be greater support and direction from governments and scientists about how to build more resilient societies. Much thought is needed on the right policies and signals that will spark the transition to a more sustainable global economy when people's livelihoods, cultures and values are at stake. Renewable energy technologies are not a panacea. In the short term, they are unlikely to be able to meet the world's energy needs. People won't switch energy sources, adopt renewable technologies or make changes to their lifestyle unless they are forced, bribed or educated into it. Merely producing new technology does not mean that it will be adopted. Politics, resources, history, psychology and social context are all important in persuading people to shift to new technologies.

Many people find it difficult to connect the concept of global climate change with their everyday actions: what's the connection between flooding in Mozambique and leaving the lights on in our homes when we go out? The news media don't help: dramatic reports of major natural events like storms, floods and droughts leave many people despondent and sceptical of the link between their small actions and huge climate change. Disaster movies, such as *The Day After Tomorrow*, reinforce this disbelief, leaving audiences either mistrustful of climate change or feeling terrified yet powerless to act.

Societies face the collective challenge of navigating the future without a clear map; to prepare for coming crises as individuals, as families and as communities; at regional, national and international levels. Not preparing for climate change will leave all societies vulnerable to its impacts. People as individuals, consumers, shareholders, producers, decision-makers, planners and voters must engage with the issues and demand that our politicians – at all scales and all levels – plan now. In whatever guise, we need to take climate change seriously and think about how to survive it. Will our actions be adequate and will we adapt quickly enough?

Actions speak louder than words

We believe we can make a reasonably accurate prediction about how climate change will affect sea levels and global temperatures. But for many other atmospheric phenomena, such as storms, winds, fog, and clouds, there is little consensus. With so much unknown, our preparations are necessarily uncertain.

We can be sure that reducing emissions will reduce the long-term effects of climate change. Therefore, we know we need to cut fossil fuel consumption. But will incentives or penalties, fuel taxes or subsidies for low-carbon technology be the best way? Business and political leaders now recognise the need for immediate research and action on climate change but there are still many uncertainties about how to create a zero-carbon future. Climate change is a global problem and needs global action but international politics make global management difficult, because countries have different ideas about the ideal point (carbon dioxide level or temperature increase) at which to stabilise the climate and by when.

Small low-lying islands and some developing countries, which are likely to experience the worst effects, want to stop climate change as much as possible and set stabilisation targets very close to today's levels (around 400ppm of carbon dioxide). The European Union (EU) has a stabilisation target of a 2°C temperature rise, based on the assumption that at greater rises, the effects of climate change will be overwhelmingly dangerous. To achieve this target, by 2050, rich countries will need to cut their greenhouse gas emissions by up to 80% of 1990 levels. Some richer nations, such as the USA, Canada and Australia, disagree with this stabilisation target, arguing that it will throw their countries into recession. This international political hot potato is being tossed about, with agreement neither on stabilisation targets nor on how to achieve them.

Initiatives led by businesses are becoming more common although there is scope for far greater levels of corporate activity to respond to climate change. Barclays Bank was the first financial institution to join the UK Emissions Trading Scheme (the forerunner to the EU Emissions Trading Scheme); in 2004, HSBC was the first bank in the world to commit to becoming carbon-neutral. Business leaders are challenging scientists: for example, Richard Branson offered US$25 million to the first person to devise a way to extract greenhouse gases from the atmosphere. A 2007 report from Citigroup listed 74 companies, in 21 industries, from 18 countries that had identified financial opportunities resulting from climate change that they could exploit. For example, John Deere, the American farm equipment company, may benefit from higher crop prices and biofuel production, and companies whose business is in energy efficiency, such as Itron, Siemens and Johnson Controls, are likely to benefit from changes in regulation and higher energy prices.

CUYAHOGA COMMUNITY COLLEGE
EASTERN CAMPUS LIBRARY

Carbon markets are increasingly seen as a suitable method of reducing carbon dioxide emissions. It is estimated that, by 2050, the global market will be worth US$500 billion. Although some are highly critical of this market, it is an opportunity for others. The UK Government has set a target of reducing the UK's carbon dioxide emissions by 60% (of 1990 emissions) by 2020, by establishing a price for carbon. It proposes to do this by investing in a range of activities, including biofuel production and carbon capture and storage. In the USA, the Clean Energy and Security Act (Waxman-Markey Cap and Trade Bill) was passed by the House of Representatives on 26 June 2009. The bill has major implications for tackling climate change; as a result, some say, the carbon market is set to grow exponentially.

Many new public–private partnerships are emerging to tackle climate change. The city of London is the world leader in carbon trading. Having started in 2005, London is now (2009) the leading centre of European emissions trading, with a market value of over £9 billion. The London Climate Partnership, co-ordinated by the Greater London Authority (GLA), is a group of more than thirty organisations, including politicians, climate scientists, developers, financiers, clinicians, environmentalists and journalists, which aims to help London prepare for the impacts of climate change. It commissions research and ensures guidance on climate change is included in all GLA policy documents; it works to raise awareness of the risks and opportunities of climate change and provides simple, accessible information for all Londoners.

There are also local and regional efforts to change behaviour. Rebecca Hoskin's successful campaign to make Modbury in Devon a plastic-bag-free town is one example. Artists are increasingly involving themselves in climate change activism. David Buckland's charitable foundation 'Cape Farewell' is a project which brings together artists, scientists and educators to raise awareness about climate change. It has commissioned five expeditions into the High Arctic that have examined the impact of climate change. International initiatives are emerging: in 2008, the American Chamber of Commerce in Vietnam hosted a conference, 'Climate Change: Innovations in Public-Private Partnerships', to identify new technologies for investment and business development.

In the US, 'Climate Leaders' is an incentive scheme run by the US Environmental Protection Agency (EPA). It is an industry–government partnership that encourages companies to develop comprehensive climate change strategies, through production of a corporation-wide inventory of greenhouse gas emissions based on a quality management system, setting aggressive reduction goals and reporting their progress annually to the EPA. In return, they receive EPA recognition as corporate environmental leaders. Mayors across America are working with the International Council on Local Environmental Initiatives-Local Governments for Sustainability programme to encourage local action on climate change. And the Climate VISION programme, a voluntary public–private partnership started in 2001, aims to encourage thirteen energy-intensive industries to reduce their greenhouse gas emissions by 18% by 2011.

It is easier to see how we can cut back on greenhouse gas emissions than invest in adaptations to climate change. As yet, there are few substantial incentives for adaptation, despite its expected importance nationally and internationally. In the UK, adaptation ranges from improving city infrastructure (such as national government investment of £700 million in new sewage systems for London by 2014) to supporting the National Farmers Union's development of strategies to enable farming communities to adapt to increasingly extreme weather.

For most countries with a coastline, the greatest climate change challenge probably lies in working out how to deal with the growing proportion of the population that lives close to the coast. Eighty per cent of UK cities are on the coast or at the mouths of its rivers; as are many of South East Asia and the Americas' biggest and most economically important cities. Should governments start moving those cities from low-lying coastal locations? Or should they defend the coastline? Is there a role for businesses? Who will bear the costs?

2

There's no predicting the weather

'Essentially, all models are wrong, but some are useful.'

George E.P. Box, 1987

Climate change *will* affect our weather. It's just not easy to be certain exactly how. Climate modelling techniques are by no means perfect and climate impact models cannot yet make precise predictions. As the climate-sceptics often point out, even the best of today's climate models still have many glitches. For some aspects of the weather, such as temperature and rainfall, scientists can give clear and specific information, but assessing the impact of climate change on others such as clouds and fog is riddled with uncertainty. Scientists use the best-available science to show what might happen to the planet if we continue to emit greenhouse gases without care, constraint or any preparation for the consequences.

Predicting climate change isn't simple. Climate models are criticised for their limited capacity to predict changes exactly and for their lack of reliable feedback between the climate system and other natural changes. While climate scientists are the first to acknowledge the limits of their models, they also acknowledge their responsibility to provide the best possible information to decision-makers and politicians. Models enable scientists to visualise and interpret the complicated interactions of the Earth's climate system and evaluate different future scenarios of societies' choices. The validity of these scenarios, looking twenty, thirty, fifty, a hundred, thousands of years into the future, is critical; from them come the estimates of greenhouse gas emissions that feed into the climate models.

We know broadly about the consequences of climate change: yet to the layperson what we know and what we don't know can often seem like guesswork. Yet it is within these uncertainties that decision-makers – and all of us – have to make our choices about how to live. We sympathise with a public that finds it difficult to know what to believe. Science can only provide so much certainty and scientists squabble over figures that they can't seem to agree on: what is a safe degree of warming? What carbon dioxide levels should we aim for to stabilise the Earth's temperature? Will carbon trading provide a solution?

A brief history of climate science

What is the human contribution to climate change? While the majority of the world's leading climate scientists would agree that climate change is human-induced, some people strongly oppose this notion, instead attributing the changes to natural phenomena, such as sunspots.

The science of climate change has a long history. It dates back to the mid-1800s, when Jean-Baptiste Joseph Fourier theorised that the atmosphere retains heat like a greenhouse – essentially coining the term 'greenhouse effect'. In 1863, the British physicist John Tyndall (after whom the Tyndall Centre for Climate Change Research is named) realised the importance of the radiative potential of carbon dioxide, that is, that the presence of carbon dioxide can change the level of radiated energy in a system. Energy from the sun enters the atmosphere; this warms the Earth and the warmth is emitted back into the atmosphere as infrared radiation. The presence of the atmosphere is vital in keeping the Earth at a reasonable temperature for life to survive. The presence of greenhouse gases, such as carbon dioxide, exerts a positive force on the energy balance by preventing some of the infrared energy being radiated back from the Earth's surface, thus warming the atmosphere. Other emissions exert a negative force – that is, they cause more energy to leave the system – which tends to cool the climate.

Shortly after Tyndall's discovery, a Swedish scientist, Svante Arrhenius, made the first calculations of the contribution of carbon dioxide to the Earth's surface temperature. Later, in 1903, he calculated that carbon dioxide added to the atmosphere by the burning of fossil fuels could raise the Earth's temperature. (Fossil fuels, such as coal and oil, are created by the decomposition of long-dead – up to 650 million years ago – living things.) Arrhenius also predicted that a doubling of carbon dioxide levels in the atmosphere would cause the global average temperature to increase by 1.5° to 4°C – highly comparable to current predictions. These early discoveries laid the foundations for our current understanding of climate change.

Understanding the fundamental limits of science

To understand the merits of predictions made from models and the uncertainty surrounding them, it is necessary to understand what science is and how it operates. Science is a process, through which facts are derived by observation, comparison, analysis, synthesis and hypothesis. Modern science, as we know it, originated in the early eighteenth century, when scholars started to take observed evidence seriously. Before then, science was the domain of authority and philosophy: the Greek philosophers, the fathers of the Church, the Bible and other religious documents were the arbiters of knowledge.

Science is often considered special for its claim that the scientific method can deliver the 'truth'. 'Facts' or 'truths' proven by the scientific method are often considered beyond dispute: this is not so. Scientists frequently overturn other scientists' findings and science never goes unchallenged: this is the nature of the scientific method. The scientific method allows insight into the individual elements that make up the bigger picture. What can remain difficult to grasp is the complexity of the many small components that add up to a larger, hopefully coherent model. Making the links between the different pieces of the puzzle is perhaps one of greatest challenges facing modern science, because each discipline – whether it is biology, physics, chemistry, engineering, computer science, psychology or any of the many other sciences that exist today – follows its own traditions, speaks its own language, works at its own scale and does not easily share its insights. To create completely descriptive climate models, we need to be able to describe interactions between people, plants, oceans, air, animals, minerals and more.This requires interdisciplinarity. Such interdisciplinarity has never been achieved and may indeed be beyond today's scientists.

Who owns climate models and what do they do?

Models are simulations of the real world; tools to help us understand how complex systems work and to predict their behaviour. Climate models are numerical representations of the Earth's climate system. They model the atmosphere, its circulation, energy inputs and outputs, and the chemical reactions that determine the concentrations of important constituents, like methane and ozone; and also the oceans, the land surface, the cryosphere (the icy bits) and the biosphere (living systems).

Models, whether statistical, conceptual or system-based, are an integral part of modern sciences, from physics to economics.

Each has a different purpose: statistical models are used to identify what is most important in an observed pattern by measuring the influence of different factors and their interactions. Statistical models give a snapshot of a particular system in a particular state, but to make predictions about changes in a system it is necessary to take a 'process-based' approach. This approach creates concepts of a system, its respective parts and their interactions from first principles, making it possible to test whether assumptions are valid and sufficient to explain the behaviour of a system. Behind the scenes, complex systems models manage many aspects of daily life: road transport analysts use transport models to predict road use, identify traffic blackspots and devise better transport management. Health risk models (a type of statistical model) are widely used in health services to determine what level of resources to invest in different areas.

Climate modelling is several orders of magnitude more complex than either transport or health service modelling. The first global climate model to combine oceanic and atmospheric processes was the General Circulation Model (GCM), created in the late 1960s in the Geophysical Fluid Dynamics Laboratory of Princeton University, in the USA. The GCM allowed scientists to share high-speed computers to create large-scale models that represent the physical processes that drive the atmosphere and oceans. The laboratory went on to create a variant of the model, the 'Geophysical Fluid Dynamics Laboratory Coupled Model'. Enhancements and upgrades have ensured that this is one of the leading climate models now used in the Fourth Assessment Report of the IPCC.

Other climate models – perhaps twenty – have been developed in the universities and meteorological offices of the world's wealthier countries. The complexity of the models means they require significant processing power and so are extremely expensive to develop and run. Some of the other well-known models used by the IPCC are the US National Center for Atmospheric Research Community Climate System Model, the Australian Commonwealth Scientific and Research Organization model, the European Centre Hamburg Model developed at the Max Planck Institute for Meteorology, the Japanese model created jointly by the University of Tokyo Center for Climate System Research and the National Institute of Environmental Studies and the model produced by the Hadley Centre, part of the UK Meteorological Office.

To get over the problem of the demands climate modelling makes on computer power, *climateprediction.net* took a novel approach. Instead of running on one huge computer, this program drew on the processing power of thousands of personal computers around the world. Computers switched on but not being used are an unused resource of computer processing capacity – people who joined in the project agreed that their computer's spare time (which may only have been a few minutes or seconds) could be used by the program. Climateprediction.net is 'the largest experiment to try and produce a forecast of the climate in the twenty-first century'.

Climate scientists everywhere continually aim to improve their climate models. Each model has different parameters and is based on different assumptions about how the parameters interact. Although there are still many challenges, over the years there has been a trend towards an increasing consensus. In some areas of climate science, there is broad agreement: although they disagree on its extent and speed, all climate models agree that as we emit more greenhouse gases, the atmosphere will warm. In others, there is significant divergence for parts of the world such as the Amazon; some models suggest it will become dryer – potentially even arid – whereas others suggest its climate will remain similar to today's.

Although all models try to break down the Earth's complex climate behaviour into simple mathematical equations, they need huge numbers – many hundreds – of equations to be solved simultaneously. And these hundreds of equations must be run many thousands of times over: as a result, very few runs are normally undertaken. The models' very complexity can hide some of the more challenging aspects of climate modelling. Communicating the importance of the models to the public, while simultaneously explaining their uncertainties, has proven extremely challenging for scientists: 'effective communication of uncertainty is notoriously difficult and the analysis of uncertainty in climate forecasts remains rudimentary' (*Nature* commentary, 1999).

Communicating uncertainty – the models

Although the uncertainty inherent in climate modelling is a complicated issue, it is necessary that we understand it, as uncertainty lies at the heart of every sceptic's argument about the rate and extent of climate change.

There are four fundamental scientific challenges to be overcome in improving the quality and capacity of the models currently used to predict climate change and guide our future actions: first, the inability of even complex mathematical models to replicate the real world in its entirety; second, that climate models contain hidden uncertainties that can skew outcomes; third, that models can only represent what we know; and finally, that models predict climate thresholds poorly.

Acknowledging the existence of uncertainty by no means reduces the seriousness or salience of the challenge of climate change. Our aim is to show that what we are at present able to predict is likely to be the minimum threat that we face. We may experience significantly greater impacts, about which we currently know nothing. The future always holds surprises that the present can't even speculate about. In some ways, Jules Verne was an uncannily accurate predictor: his book *Paris in the 20th Century*, written in 1863, predicted air conditioning, the Internet, television and fax machines, among much else; yet in others, he was wildly erroneous: in *20,000 Leagues Under the Sea* (1870) he predicted the discovery of the lost city of Atlantis at the bottom of the ocean.

Complex mathematical models cannot replicate all aspects of the real world

Climate models are mathematical simulations of long-term atmospheric, oceanic and geographic conditions. However, even the best models are built from our current knowledge of physics and chemistry and do not yet incorporate complex biological behaviours. It is often not possible to predict how a population will behave based on the behaviour of individuals, just as it is not possible to predict the behaviour of an individual from the behaviour of its cells, or their behaviour from the behaviour of their constituent molecules.

Some climatic behaviours will always be hard for models to reproduce. For example, imposing stresses such as increasing temperature or changing wind patterns on a model will not produce a consistent response. Another difficulty is the problem of modelling random events, such as volcanic eruptions. To model a climate as a whole, without having to account for every single property, connection, interaction and feedback mechanism, climate scientists use an approach known as 'system identification' or 'semi-physical modelling' (sometimes called the 'grey-box' approach). Nonetheless, such models do need to include descriptions of processes that we do not yet completely understand – known as 'black-box' modelling. These black-box models sometimes make sweeping (and possibly erroneous) assumptions about some relationships but they must be included for as great a completeness as possible. A hybrid model, derived from both the statistical analysis of empirical data and from first principles, can describe the behaviour of a system with reasonable accuracy, even though it does not have every single 'real world' detail within it.

Our economic and social future is, as yet, notoriously difficult to predict. How climate change will affect us depends on the extent of future greenhouse gas emissions, our capacity to render those gases harmless and our ability to prepare for and cope with the impacts of change. Yet the picture is complicated by thousands of other factors, from future population distribution, global income patterns, use of natural resources and types of government to changes in technologies.

Climate models contain hidden uncertainties that can skew outcomes

There are many 'unknowns' in climate models: it is not possible either to measure their effect or know how they will affect the outcomes of the model. To have reasonable confidence in our models, we must understand how their predictions are affected by the uncertainty of these unknown parameters and processes.

There are four main sources of uncertainty: first, intrinsic uncertainty – the variation in the elements included in the existing climate models; second, extrinsic uncertainty – the uncertainties in socio-economic scenarios and the prediction of human behaviours; third, non-linear dynamics, when feedback is not taken into account; and finally, regional uncertainties that occur when global models are scaled down to finer resolutions. However, despite these uncertainties and variations, it is worth noting that all models agree in predicting atmospheric temperature increases of between 1°C and 10°C by 2100 if behaviours and conditions remain as they are today.

Models can only represent what we know

We do not yet understand some physical processes well enough to incorporate them reliably in climate models. While, given time, our ability to model the climate system will improve and our ability to identify how the climate will respond to different levels of emissions will become clearer, present models, while mathematically complex, are nonetheless quite basic in what they include.

Importantly, they do not yet model very complex processes, such as non-linear dynamics, physico-chemical feedbacks and local conditions. Non-linear dynamics that are presently poorly modelled include the die back of the Amazon due to changes in the balance between rainfall and evaporation (transpiration) of water from plants and alterations in the composition of plant communities, for example increases in numbers of 'C4 plants' – plants such as sugar cane and maize which are able to absorb carbon dioxide (used in photosynthesis) more efficiently than other plants.

Physico-chemical feedback refers to the influence of one process or system on another. For example, higher temperatures could lead to the drying-out of peat bogs, which could release a significant amount of carbon dioxide, which could rapidly add to the global load of this greenhouse gas. Other physical-chemical reactions include the destabilisation of gas hydrates and the melting of tundra and glaciers. Gas (or clathrate) hydrates are crystalline solids, found on the deep ocean floor, in which molecules of gases such as methane, carbon dioxide and oxygen are trapped inside a lattice of water molecules – their breakdown could lead to the release of the trapped gases. The melting of frozen arctic tundra could also lead to a massive release of methane. Melting and reduction of glaciers could lead to reduced reflection of infrared radiation to the atmosphere, leading to a faster rate of global warming.

Local conditions are very difficult to model. While most global climate models agree on global and regional trends, they do not accurately predict local conditions. For example, they are poor at predicting extreme events; when they will occur, how often they will happen or their severity. Extreme events, such as violent storms, severe floods and lingering droughts are part of the normal climate: we expect them, but our models are not very good at predicting exactly where or when they will happen. We do not know how climate change will affect such events – will they become more or less intense; more or less common? Some of the less well-understood aspects of the Earth's complex climate system, such as the relationship between the West African monsoon and hurricane activity in the Caribbean, could be profoundly affected by climate change – yet we do not know how.

Models do not predict thresholds of climate systems

Missing details, hidden uncertainties and unknown factors in climate models mean it is difficult to predict thresholds in climatic systems. At a threshold, the climate tips from one stable state to another – different – stable state. Present models find it hard to pinpoint reliably when this change will occur. For example, present agricultural and social patterns in the UK depend heavily on the stable circulation of the warm Gulf Stream through the North Atlantic; however, even tiny changes in ocean salinity or temperature (caused by global warming) could lead to huge changes in the direction and flow of this current, which would radically alter the UK's climate. Presently, the UK has a stable, mild climate; if the threshold is passed, we could move to a stable, colder climate, akin to Norway's. Climate models vary in their ability to predict both if a threshold will occur and the results of it.

Communicating uncertainty – the practical challenges

Although the levels of uncertainty in climate models are very high, they are nonetheless the best tools we have in predicting climate change. Scientists face the considerable challenge of communicating this uncertainty – exactly what we know, what we do not know, what we are likely to know in the next few years and what we will probably never know – without making their explanations too simple and likely to hide the reality of climate change. How can we illuminate these issues and communicate the truth without producing confusion and leading some people to believe that climate change is not happening?

The surface temperature of the Earth is increasing and rainfall patterns are changing. These seem to be impersonal concepts, unrelated to our real lives. What we really want to know is: will my town be flooded? Will there be a drought this summer? Are we in for a stormy winter? Are the cliffs eroding more quickly than they used to? Is there going to be a heatwave this summer? If people have the information to answer questions like these, they are more likely to engage with the issues behind climate change and become aware of the probabilities and risks of its various impacts.

Climate surprises

Although today's climate models are imperfect, they still enable us, when used with care, to take the best actions possible to reduce future harm. One approach is to focus on reducing the vulnerability of those already being harmed by climate change. This should be good enough to enable us to cope with the dangers associated with 'normal' climate change; however, we need to be aware of the possibility that the climate may cross a threshold and 'flip'. Climate scientists disagree on how likely such an event might be – many argue that we are a very long way off but many argue that it could happen very soon, if it hasn't already.

There is also the risk of potentially catastrophic 'rapid' or 'abrupt' climate change. Although there is no agreed definition, rapid, in this context, refers to impacts greater than those we can expect from normal climate change. Most scientists draw a broad distinction between two types of rapid climate change: sudden changes, in which the climate system crosses a threshold and flips to a new state; and accelerated change, in which processes are speeded up.

Sudden changes include the collapse of the Thermohaline Circulation (large-scale ocean currents) and the collapse of the Greenland and West Antarctic ice sheets. The Thermohaline Circulation brings warm water across the North Atlantic, leading to relatively mild climates in Western Europe (see Figure 4). Without it, temperatures across western Europe would probably be up to 3°C lower than they are now. The last time temperatures dropped this much was about 12,000 years ago, in the Pleistocene epoch (at the end of the most recent major Ice Age, long before human civilisation began). At this time, despite the Earth generally being in a warming phase, over about a hundred years, cold climate conditions returned which lasted for another 1,300 years (a climatological period called the Younger Dryas). There is still debate among scientists about the exact cause, although many speculate that it was the result of a partial or total shutdown of the Thermohaline Circulation. In December 2005, researchers led by Harry Bryden, of Southampton University, UK, reported in *Nature* that they had identified a weakening of the Thermohaline Circulation by about 30% since the early 1990s. Current climate models now simulate a future weakening of the Thermohaline Circulation. However, its complete collapse is considered likely only under extreme conditions.

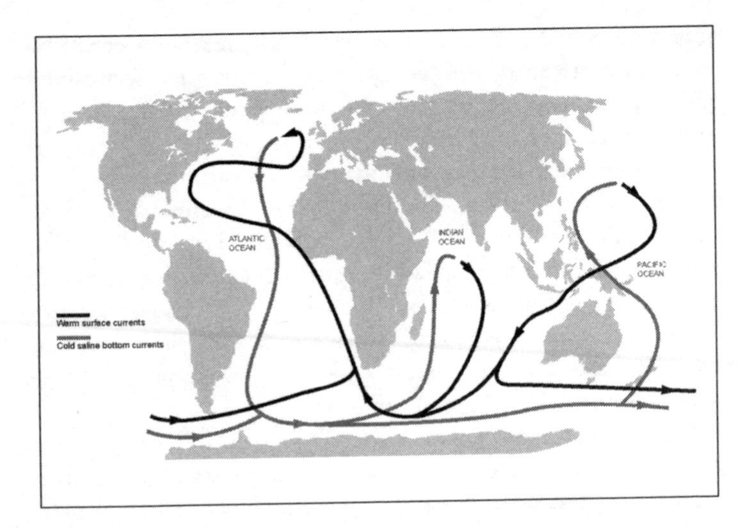

Figure 4 The Thermohaline Circulation (source: adapted from Figure 4–2 IPCC (2001) *Climate Change 2001: Synthesis Report.* Intergovernmental Panel on Climate Change, Cambridge University Press, Cambridge)

Research reported in 2006 by Luthcke et al. confirmed that the Greenland ice sheet is melting at an accelerated rate. Current estimates claim that it is losing about 100 gigatonnes of ice every year, compared to losses of about twelve gigatonnes per year between 1992 and 2002. The West Antarctic ice sheet is also melting but changes in the Antarctic have been so rapid that scientists are finding it difficult to keep up with the changes and account for them in their models. A huge volume of ice is stored in the Antarctic; a partial melting could result in a sea level rise of four to six metres across the globe, although this could take hundreds or thousands of years to occur. A complete melting is less likely but would lead to sea level rises of five to seven metres.

Accelerated change refers to processes that are speeded up as a result of climate change. Most often, this type of change is due to positive feedback loops within the climate system. Positive feedback enhances the output of a system; the end result of positive feedback is an amplification – a small push results in big changes. For example, as the atmosphere warms it warms the Earth and water vapour is released from frozen Arctic tundra, peat bogs and other wetlands. Water vapour being a very powerful greenhouse gas, this increases warming of the atmosphere. A warmer atmosphere can store more water vapour, which increases the speed of warming, and so the cycle goes on. A second example occurs as sea ice begins to melt – with less sea ice, less solar radiation is reflected back into the atmosphere, so is absorbed by the ocean and heats it further. Yet other positive feedback loops have the potential to produce substantially larger increases in temperature than currently projected. Many examples of positive feedback loops are neither well-understood nor adequately modelled; scientists believe accelerated change may bring about an abrupt change in climate but it is unlikely.

For all that the possibilities of rapid climate change have captured the media and public imagination, through films such as *The Day After Tomorrow*, scientists are in little agreement on the likelihood or timing of rapid climate change. However, there is a general consensus that there is a small chance that rapid changes could occur in the future; after all, abrupt climate shifts have occurred in the past (as in the Younger Dryas). Almost no research has been undertaken to assess what can be done to prepare society for these unlikely, but potentially cataclysmic, shifts in climate. Even so, such changes should be considered dangerous, because they are difficult or impossible to adapt to, and are likely to involve significant costs and produce irreversible changes.

3

How climate change will affect our lives

> 'Global atmospheric concentrations of carbon dioxide, methane and nitrous oxide have increased markedly as a result of human activities since 1750 and now far exceed pre-industrial values.'
>
> IPCC Fourth Assessment Report, 2007

For the first time since the dawn of the human race, humankind has the means to radically affect the global climate system. Our production of greenhouse gases is causing climate changes that are unlikely to be benign. Many changes are likely to happen as a consequence of these emissions; we intend to focus only on those that – given the current state of knowledge – scientists consider are very likely to occur. These are the changes that we can start to prepare for. However, there will be many more impacts that we do not yet know about; hopefully, the forecasting skill of the models, which will enable us to identify these new impacts, will improve more quickly than the speed at which climate change unfolds.

Is the outlook for our planet irrevocably bleak? The short answer is 'no!' Climate change will not necessarily harm all living things or their environments. The extent to which they are affected will depend critically on three things: how much we are able to reduce our emissions of greenhouse gases, how we prepare for changes associated with climate change and how we respond to climate changes as they occur.

Vulnerability to any hazard is determined by exposure,

sensitivity and the adaptive capacity of the components of the system. 'Exposure' relates to the frequency and magnitude of the hazard, often determined by geophysical factors, such as the topography of a landscape or the extent of tides. 'Sensitivity' is

BEST PRACTICE FLOOD RISK MANAGEMENT

In 2000/2001, two groundwater floods affected the parish of Hambledon in Hampshire, UK. After the floods, the Parish Council drew up a checklist of key issues to be aware of in future; their work is an example of good practice, in which the Parish Council worked with the national Environment Agency to manage flooding.
Key issues:

Know the threat
Know the likely flood interval
Know the responsibilities for floodwater management
Know the floodwater management plan
Establish a flood warning system
Establish a community communication system
Nominate a flood action co-ordinator
Know the importance of self-help and be willing to learn
Establish a 'flood information centre'
Keep the local authorities informed
Keep accurate records
Control the traffic
Keep businesses open
Obtain the necessary manpower
Take care of the elderly
Check public health and security
Look after the media
Call village consultative meetings
Take tidy-up action
Capture the lessons
Prepare a flood emergency plan

Source: www.environment-agency.gov.uk

the way in which a hazard is experienced, and is influenced by social and political factors (such as whether building regulations account for the risk of flooding and, indeed, whether they are implemented). 'Adaptive capacity' refers to the way in which living things adapt to cope with changes in their environment; from rabbits digging a deeper burrow to cope with higher soil temperatures to people putting aside savings 'for a rainy day'.

Reducing sensitivity to exposure is a well-established way of managing the risk of hazards. This can be seen in the way in which flood risk is managed in the UK. Floods are managed at many levels: national, regional, city and parish; the plans developed depend on where action is taken. Preparedness occurs at all levels and through many types of planning: the national government's plans are strategic and address the big issues of security, public health and resource management. Local authorities' plans are different again. And people make their own plans to face the risks of where they live. For example, forest fires can damage buildings and threaten life, so people have good reason to take action to reduce their vulnerability. Most people living in areas where forest fires are frequent know the risks and make sure they have prepared for them. Homes can be built using fire-resistant materials, combustible materials can be stored some distance away, buildings can be sited away from combustible plants and dry grass and excess vegetation can be removed. Forest residents can draw up evacuation plans: what is the safest route to take away from the fire, what important items should they take with them, what resources will they need to survive? Such plans increase people's ability to cope with disasters affecting their homes; similar planning could enable humanity to cope with climate change.

However, even the very best planning cannot necessarily help us cope with impacts of climate change which may be very different to those we are experiencing now. So, governments are incorporating adaptations to climate change into their planning for disaster risk reduction and – as far as they can foresee – disaster-proofing their strategies. In northern Bangladesh, one of the effects of climate change is likely to be more frequent and more intense river floods. These floods are particularly dangerous for the communities who live and grow their food on the many mobile sandbanks – 'chars' – within the rivers. Char-dwellers know that they need to build on the highest ground possible; existing risk reduction measures include creating village disaster management committees and helping homeowners raise the level of the ground on which their homes and kitchen gardens are built. Climate change means that additional measures will be needed, such as creating rice seed banks so that rice can be replanted quickly if the harvest is lost.

Broad impacts

Preparing for climate change, mitigating and adapting to its effects and coping with its impacts will change the way in which we experience climate change. But what might happen if we don't take any active measures?

Rising temperatures

In some regions, a slightly warmer climate doesn't seem like too much to worry about. A few extra degrees on the thermometer could produce warmer summers in the Earth's mid-latitudes (the regions between the tropics and the poles, 30° to 60° either side of the Equator). In the low latitudes (the region between the Tropic of Cancer in the north and the Tropic of Capricorn in the south), summer temperatures frequently exceed 38°C but significantly higher temperatures have been recorded. In both Death Valley in California and El Azizia in Libya, temperatures of up to 58°C have been registered. Many people ask, if temperatures in some areas can normally vary by up to 20°C, can one or two degrees of warming really make that much difference?

Of course, local heating and cooling are very different from the warming of the atmosphere around the entire planet. To assess the impacts of global (rather than local) warming, we must take into account the effects of past global warming. In Chapter 1, we looked back at the impact of the 0.75°C warming that has taken place since the mid-1800s. Considered in this light, two degrees of warming could make the planet a very different place.

Heatwaves

Climate change is likely to produce more intense heatwaves with higher peak temperatures that are more frequent and longer-lasting. The severe heatwave experienced in Europe over twenty days in August 2003 has been held responsible for many deaths: it is estimated that in France alone approximately 14,802 'extra' deaths occurred, compared to recent years. Although many things contributed to this high death toll – not least that August is the traditional French holiday period and many Parisians were away from the city, leaving their elderly relatives at home (the majority of those who died were elderly, ill, terminally ill or in some way unable to care for themselves) – undoubtedly the great heat was a major contributory factor.

Rising sea levels

Of all the possible effects of global warming, none are likely to be as immediate, as unstoppable or as significant as the impact of rising temperatures on the oceans. When water is heated it expands: a warmer atmosphere heats up the oceans; consequently the oceans expand more. This 'thermal expansion' of the oceans accounts for 70–75% of sea level rise – the melting of glaciers, ice caps and the Greenland ice sheet also contribute.

Inundation caused by higher sea levels is a major hazard for low-lying islands and coastal communities around the world. The IPCC 2007 report estimated that sea levels will increase by 0.18m to 0.59m by 2100. The rise will not be the same everywhere, due to the unevenness of the seabed, the impact of gravity on the oceans and the tectonic movements that push some areas of the Earth's crust up and others down. Stefan Rahmstorf (writing in *Science* in 2007) experimentally incorporated various physical processes (including thermal expansion of oceans, melt-water from glaciers and ice sheets and changed water storage on land) into a coupled ocean–atmosphere model. His findings suggest that by 2100, sea levels are likely to be between 0.50m and 1.40m above the 1990 level. However, he noted that if the ice sheets melt completely (a phenomenon as yet poorly understood and poorly modelled), global sea levels could rise by as much as 70m.

Ocean acidity

The increased concentration of carbon dioxide in the atmosphere will also change the composition of the oceans; more dissolved carbon dioxide makes them more acidic. At the time of writing (2009), very little is known about the broader impacts of an increasingly acidic ocean.

Naturally, the oceans are predominantly alkaline, which is a necessary condition for organisms such as corals and algae to extract calcium from seawater and deposit it as solid calcium carbonate rock. More acidic water is likely to dissolve these carbonate sediments; critically affecting coral reefs and the fish that rely on them for their homes and food. Coral reefs, which provide a degree of buffering against wave erosion, are also at risk from higher ocean temperatures.

Droughts and floods

It seems paradoxical that climate change could bring both more flooding and more droughts, yet that is exactly what climate models predict and what we are starting to see. Climate predictions show that rainfall will increase, especially in monsoon areas; the tropical Pacific, in particular, can expect heavier rainfalls during the rainy season. Together with this change in the intensity of rainfall, there is likely to be a change in the annual distribution of rainfall – for example, the UK can expect to see more rainfall during the autumn and winter but less in the summer, which has implications for the provision of an adequate water supply. It also implies that we can expect more summer droughts and autumnal floods.

Tropical storms

'Tropical cyclone' is the generic name for the circular, windy and rainy storms prevalent in the tropics (called hurricanes in the Caribbean and typhoons in the northern Pacific). There is still much debate among scientists about the exact nature of the relationship between climate change and tropical cyclones but it is largely agreed that intense tropical cyclones are likely to become even stronger, while weaker storms become less frequent. There is also speculation that cyclone tracks may alter, so that people and places previously unaffected by them, and with no traditional understanding of how to cope, may start to experience their havoc.

Climate change around the world

It is important to assess how climate change will affect different parts of the world. Not all regions will experience it in the same way, because climates vary around the world and different levels of preparation will be made. Recent advances in regional climate modelling mean that we have a much better idea of how the climate will change in different parts of the world.

Africa

'Heat' and 'drought' are the likely descriptors of Africa's future climate. The impacts of these climatic changes are likely to be severe: there is general agreement that Africa is the region most vulnerable to climate change and variability, due to its existing vulnerability.

Africa is very likely to warm relatively more than other parts of the planet and experience higher average temperatures throughout the year, with subtropical areas warming more than tropical areas. Around the Mediterranean and in both southern and northern Africa, rainfall is likely to decrease. However, rainfall is expected to increase in eastern Africa. We do not yet know how rainfall will change in the Sahel and the southern Sahara.

Seventy-five to 250 million more people are likely to suffer water shortages. Water scarcity will have a profound impact on African agricultural production, as the area suitable for agricultural production is reduced. By 2020, we can expect that yields from rain-fed agriculture will reduce by 50%. Rising water temperatures will put pressure on fish stocks in large lakes; these resources are already under stress from over-fishing.

Asia

Severe rainfall, increasingly intense tropical cyclones and more frequent, longer and more extreme heatwaves could describe Asia's future climate. In central Asia, the Tibetan plateau and northern Asia, warming is expected to be much higher than the global average; the rest of Asia is expected to warm at about the global average rate. Winter precipitation is very likely to increase in northern and eastern Asia and the southern parts of south-east Asia. Summer rainfall is expected to increase throughout Asia, except in central Asia, where it is expected to decrease.

As a result, the Himalayan glaciers are likely to melt more quickly, potentially causing catastrophic flooding and rock avalanches on unstable slopes. As the glaciers retreat, river flows will decrease, leading to water supply problems. By 2050, the combination of climate change and population growth is likely to create freshwater shortages in central, southern, eastern and south-eastern Asia, potentially affecting more than one billion people. Regional climatic variations mean that crop yields could increase in some areas while decreasing in others. The overall impact on food production is hard to predict, although clearly, those areas where yields decrease will face problems of hunger.

Asia's many heavily populated, extensive river deltas are at risk of flooding. More flooding means more water-borne diseases: diarrhoeal disease will remain a significant killer and there will be an increasing risk of highly toxic forms of cholera, associated with warm coastal waters. The biggest challenge for Asia will be to find sustainable development pathways that allow continued urbanisation and economic development within the constraints of natural resources already under pressure from climate change.

Australia and New Zealand

Water security, coastal storms and loss of coastal biodiversity will be key challenges for Australia and New Zealand. Drought is a particular threat, especially in southern Australia. Australia and the northern part of New Zealand will probably warm at about the global average, with more days of extremely high temperatures. The southern parts of New Zealand are likely to warm relatively below the global average rate. Winter rainfall is very likely to be lower in southern Australia but higher in the west of New Zealand. And when it does rain, it is likely to be heavier everywhere, except in southern Australia in winter and spring. Average wind speeds are expected to increase in the South Island of New Zealand.

Coastal residents will face an increased risk of flooding and associated economic disruption. Agriculture in New Zealand will show higher yields, because of less frost and longer growing seasons but by 2030, drought and flood will cause decreases in production over much of southern and eastern Australia and parts of New Zealand. The well-developed economies of these countries, and their scientific and technical capability, will enhance their ability to adapt but the capacity of the natural environment to adapt may be limited.

Europe

Heatwaves, droughts, wildfires and flash floods will become relatively common if climate change continues to unfold in its current pattern. Temperatures in Europe are expected to rise relatively more than the global average. Northern Europe can expect to see its greatest warming happening in winter, while southern Europe will experience the greatest warming in summer. Annual precipitation is expected to increase in northern Europe; the snow season will be shorter and there are likely to be more heavy rainstorms. In contrast, annual precipitation, as well as the number of rainy days, is likely to decline in southern Europe. There is a strong likelihood of summer drought in central and southern Europe. Higher temperatures and more droughts will affect hydro-power, crop productivity, tourism and public health. Coastal regions will experience worse flooding and erosion; mountainous areas will see glaciers retreat and snow cover reduce; inland areas can expect more flash floods.

For northern Europe, there will be some short- to medium-term benefits from climate change, such as less need for installing heating and increased crop yields and forest growth, but these will be outweighed in the longer term by the hazards of winter flooding and its associated ground instability.

Central and South America

Changing rain patterns are likely to be the biggest problem for Central and South America. Annual rainfall is expected to decrease in most of Central America and in the southern Andes, although there could be very large local variations in the mountains. Winter rainfall is expected to increase in the southern tip of South America and summer rainfall in the south-east of South America. It is not yet known how rainfall will change over the northern parts of South America, in particular over the Amazon, northern Brazil, Ecuador and Peru. This region is expected to warm at about the global average rate.

The risk to biodiversity is severe: changing rainfall patterns will affect vegetation throughout the region. In eastern Amazonia, tropical forest could become savannah, and semi-arid vegetation arid land. In already dry areas, agricultural land will be affected by salination (increased salt levels) and desertification, reducing crop and livestock productivity, with consequences for food security.

The combination of glacial retreat and changes in precipitation patterns is expected significantly to affect the availability of water for human consumption, energy generation and agriculture. While some countries have invested in early warning systems, flood (and drought) management and risk management, these efforts have been offset by a lack of basic information, low income and widespread settlement in vulnerable areas.

North America

More, stronger and longer heatwaves, more floods and more droughts characterise the future climate of North America. Overall, North America has limited readiness for the coming changes.

Most of North America is expected to warm by relatively more than the global average. Warming is likely to be greatest in winter in the northern regions and in summer in the south-west. Minimum winter temperatures are expected to increase, as are maximum summer temperatures. Rainfall is expected to increase in the north (especially Canada and north-east USA) and decrease in the south-west. The snow season is expected to shorten and snow depth to reduce, except in the northernmost parts of Canada, where snow depth is expected to increase.

Water is of paramount concern to North America; more winter flooding and reduced summer flows will increase competition for limited water supplies. The risk of forest fire will be greater, as will the area of forest affected by fires. There are likely to be some short- to medium-term benefits; rain-fed agricultural yields will increase by between five and twenty per cent but crops that rely on high water use or can only be grown in a narrow temperature range will suffer. In coastal areas, the pressure from sea level rises, combined with human activity, will create difficulties.

The polar regions

Warming is the biggest threat to the polar regions and nowhere is the warming likely to be felt as quickly or as severely as the Arctic, which is very likely to warm relatively more than the global average. Annual precipitation is expected to increase and Arctic sea ice to decrease in extent and thickness. The Antarctic is also expected to warm, with higher precipitation levels. Little is yet known about the frequency of precipitation and days of extreme temperature in the polar regions.

Losses of ecosystems and habitats, and reduced numbers of migratory bird species, mammals and higher predators (such as polar bears) can be expected due to the reduced thickness and extent of glaciers and sea ice. New invasive species will colonise the changed habitats. Traditional ways of life for human communities are at risk but there are some benefits, such as reduced heating costs and more navigable northern sea routes.

Adaptation is already occurring in the Arctic. For example, Polar View (part of the European Space Agency's Global Monitoring for Environment and Security initiative) provides satellite information on snow and ice conditions. This information is chiefly used by sailors to navigate in the Arctic but also by Inuit hunters, who use the weekly maps to determine the location of the ice's edge and improve the safety of their hunts. Despite these advances, there may be physical and psychological limits to adaptation.

Small islands

For the low-lying islands of the Caribbean Sea, Indian Ocean and Pacific Ocean, the biggest threats from climate change are rising sea levels and extreme weather. Islands can expect more intense rainfall associated with tropical storms, although more investigation is needed to assess whether there will be any change in the intensity or frequency of storms. The warming on these islands is likely to be relatively less than the global average. All islands are expected to experience changed rainfall patterns, although little detail is yet known.

Higher sea levels will cause increased erosion, worse coastal inundation and more damaging storm surges, damaging coastal infrastructure and settlements. Groundwater sources may become contaminated with salt, which will exacerbate the water stresses caused by changing rainfall patterns. Islands, especially in the mid and high latitudes, are also likely to be increasingly prone to invasion by non-native species.

To prepare for climatic changes in the Caribbean, the Canadian International Development Agency funded the Caribbean Planning for Adaptation to Climate Change project. Its purpose was to monitor change in the Caribbean and to prepare the island nations. Between 1997 and 2001, eighteen sea level and climate monitoring systems were installed in twelve countries. The information they generated is enabling governments to make better-informed decisions.

Unifying patterns

We know that the Earth's climate will change in diverse ways. It is critical to assess exactly how these changes will affect our societies, our economies and the way we live our lives.

The world will warm by anywhere between 1.1°C and 6.4°C by 2100, depending on the extent of greenhouse gas emissions in coming years. Warming within this range is the most likely – but not the only possible – outcome of climate change. There are many unknown factors, whose effects we cannot predict. Nor do we know whether a certain level of warming (perhaps just 2 or 3°C) will tip us over the threshold where normal climate change becomes rapid climate change. We do know that more heat energy trapped in the Earth's atmosphere means more evaporation of water, leading to higher levels of water vapour in the atmosphere. This will lead to higher precipitation in and around the tropics and in the sub-polar and polar regions, but lower precipitation in the subtropics.

One of the challenges of climate change is attribution: how can we be sure that the changes we are seeing should be attributed to climate change and not to other causes? If carriers of disease (mosquitoes, rats, fleas, ticks, bats) are suddenly found in a new location, can this justly be attributed to a changing climate? Or might it be due to changing patterns of human behaviour (holidaying foreigners accidentally taking the disease-carriers home with them), changing micro-climates or changing trading partners (disease-carriers being transported along with imported and exported goods)? Even though we cannot attribute every event or every change to climate change, we can identify the sectors and systems on which the climate may have an impact.

The 2007 chikungunya epidemic in Italy provides a good example. The chikungunya virus is carried by the Asian tiger mosquito, *Aedes albopictus*, a native of south-east Asia. No cases of this virus, and no *Aedes albopictus*, had been seen in Italy before June 2007, when the disease was first identified in a visitor from India. By July, there was evidence of contagion and by September, 264 cases had occurred. Should this outbreak be blamed on tourism, travel within the country or the invasion of *Aedes albopictus* into Italy? It is hard to discern a clear signal to link this outbreak to climate change. Nonetheless, epidemiologists argue that such events are typical of the type of outbreaks that could be caused by climate change.

Changes in the amount of rainfall and its distribution will make water management more difficult. Based solely on our capacity to manage the floods, droughts and fires we experience today, it is clear that we have a lot of work ahead of us to be able to cope with some of the very significant changes that are coming. Floods at one point in the year; drought at another – water, river and sewage managers will struggle to adapt an infrastructure built to withstand today's climate to cope with a shifting climate and its associated changes.

At high latitudes, and in some wet tropical areas, river run-off is expected to increase by ten to forty per cent. Such increased flows could produce significant flooding, especially where there has been extensive development on flood plains or massive human modification of natural river systems. In the dry tropics and the mid-latitudes, rainfall could be ten to thirty per cent lower, putting additional stress on areas that already suffer from low water flow. This is certain to adversely affect the livelihoods of many who rely on rain-fed agriculture.

THE 2007 FLOODS IN CENTRAL ENGLAND

June and July 2007 were the wettest on record; 390mm of rain fell on England and Wales. Three rivers (the Severn, Avon and Thames) burst their banks; the Environment Agency described the subsequent events as the 'worst floods in modern times'.

Compared to the 1947 floods (the previous worst floods) the floods of 2007 were more intense, yet caused less property damage. Tens of thousands of buildings lost electricity and water supplies but only 45,000 properties were damaged, as opposed to 100,000 in 1947. Repair costs were also lower: the 2007 costs were put at £2.5 billion; the 1947 costs at around £3 to 4.5 billion. This may be due to effective flood management by the Environment Agency, together with improved 'hard' flood defences such as better permanent and temporary flood barriers and bigger sewers. And the reductions must be set against the UK's increasing exposure to flooding: despite a long history of flooding, since 1947, at least 25% more land has been built on, including previous 'soft' defences, such as flood plains.

Ecosystem survival

Ecosystems provide us with the food we eat, the fibres that make our clothes, fuel for construction, energy and manufacturing, bio-chemicals and resources for agriculture, pharmaceuticals, medicines and cosmetics. Ecosystems regulate water purification, waste treatment, local climate, disease transmission, natural hazards and soil erosion. Ecosystems are the life-support system of the planet, an integral part of the complex cycles of agriculture, pollination, diseases and pest control. And ecosystems provide us with beautiful places and spaces for our recreation. The expected droughts, heatwaves, floods, ocean acidification and temperature will wreak huge changes on many of the ecosystems on which we rely. Twenty to thirty per cent of species could be at risk of extinction if temperatures rise by between 1.5°C and 2.5°C.

We are not yet entirely sure how climate change will affect our ability to cultivate land, feed livestock and ultimately ourselves, but it undoubtedly will. Higher levels of carbon dioxide may enhance productivity in some crop plants (notably wheat, rice and soybean; staple foods of the mid-latitudes); more carbon dioxide means more photosynthesis and more carbohydrate production. However, plants of the low latitudes, especially C4 plants such as maize, sorghum, sugar cane, millet and many grasses, thrive at lower levels of carbon dioxide and so may not grow as quickly.

Growing seasons (especially in the more northern latitudes) may lengthen by up to ten days for every 1°C rise in temperature. The geographical ranges of crops will move north and south as climate zones shift. In a 1990 report, the United Nations estimated that, in the mid-latitude regions, crops will shift poleward by about two to three hundred kilometres for every degree of warming. However, the report also suggests that while the lands at the present limit of the mid-latitude agricultural zones (northern Canada, Scandinavia, Russia and Japan in the northern hemisphere and southern Chile and Argentina in the southern) may benefit from higher temperatures and carbon dioxide levels, the soil may not be suitable for agriculture.

Soil moisture will decrease as evaporation increases; typically, in the mid-latitudes evaporation rates will increase by five per cent for each 1°C rise in temperature. Greater soil dryness could reduce crop yields by ten to thirty per cent. Extended heatwaves will have serious effects on certain subsistence and cash crops, completely counteracting the benefits from the higher levels of productivity associated with higher carbon dioxide levels. In the low latitudes, lower rainfall and soil moisture will affect crops in semi-arid regions. And hotter weather will make conditions more difficult for livestock in humid tropical regions.

About half of the world's population (about 3.2 billion people) lives within two hundred kilometres of a coast. Coastal areas are vulnerable; prone to erosion, which is likely to be worsened by sea level rises, and subject to storms, which are likely to become more frequent. Sea level rises may also cause the loss of coastal wetlands – which provide protection from flooding – when there is no available space for them to migrate inland. The low-lying, extensive river deltas of Asia and Africa, already prone to flooding and tropical storms and home to many people with a relatively low capacity to adapt to change, are particularly vulnerable. For coastal cities, towns and industries, the severity of the effects of climate change is likely to be significantly higher, due to their dense human populations.

Health and mortality

Heatwaves, floods, storms, fires and droughts already cause death, disease and injury; climate change will only make this worse. Higher concentrations of ground level ozone could increase the incidence of cardiac and respiratory diseases; lack of clean drinking water could increase the frequency and spread of diarrhoeal diseases; increased droughts will exacerbate malnutrition and its associated disorders, with implications for child growth and development; and disease-carriers will shift their geographic range, bringing novel diseases to a vulnerable population. And although, in the mid-latitudes, warmer temperatures could reduce mortality linked to extreme cold, heat-related mortality in Europe is expected to worsen, as is the incidence of pollen-related allergies. Overall, the positive benefits of warmer temperature are expected to be significantly outweighed by negative impacts, especially in developing countries.

Effects on societies

Since the late 1990s, scientists have put much effort into finding out how different societies will be affected by climate change. It has become very clear that a society's vulnerability is influenced by many factors, all of which need to be taken into account when trying to estimate the effects of climate change. Fundamentally, those who are already vulnerable to existing social, economic and environmental stresses are likely to feel the effects of climate change strongly.

It is not just physical stress that makes people vulnerable but also other local, regional and global pressures. Increasing globalisation is one such. 'Globalisation' is the removal of trade barriers between nations and the opening up of regional and local economies to the free movement of trade, money, capital, people and commodities. Globalisation's advocates argue that it will allow those who are able to make the cheapest products – often those with the lowest wages and lowest material costs – to sell their products across the world, increasing their share of the market, sales volume and revenue, and boosting their income. Anti-globalisation protestors disagree, pointing out that globalisation is inherently unfair: those already in control of global markets use a variety of practices to make it difficult for new people and companies to enter a market, so that existing multinational corporations extend their global reach and squeeze out smaller, local, competitors.

Some groups are attempting to engage with world markets as a strategy for escaping poverty. However, this can be both a lifeline and a millstone, as prices for commodities are beyond the control of local producers, environments and markets. International fashions and fads, foreign market fluctuations, global weather patterns and the behaviour of competitors in other countries are all aspects of globalisation. The current rise in levels of economic participation of newly-industrialised countries (notably China, Brazil and India) means producers in poorer parts of the world are likely to become more, not less, vulnerable to the price and demand fluctuations of global markets.

Many other local and regional factors can expose individuals to greater stress or make them more sensitive to the impacts of that stress. Poverty, disease, social exclusion and disadvantage, poor living conditions and exposure to natural or man-made hazards all increase individuals' vulnerability. This vulnerability can be reduced through greater government regulation, poverty eradication strategies and practical strategies, such as preventing people from living in risky areas like eroding coastlines or regions prone to landslides or subsidence.

The effects of climate change will not only increase the number of people who are vulnerable but also a larger proportion of the population. Geo-physical conditions affect people's vulnerability to risk; those living in low-lying river deltas may experience sea or river flooding, those living in the mountains may be more prone to landslides and rock falls, those living in flood plains and those living on small islands are at risk from flooding. Not everyone exposed to danger or hazards will experience them in the same way: social factors and government policy also affect vulnerability.

Some governments, aware that people will suffer from the effects of climate change in their current homes, have gone so far as to help them to migrate. Other governments have chosen to allow people to fend for themselves. The UK government has quietly introduced a policy of 'managed coastal realignment' in some areas: this translates to 'accepting that the coast is eroding and allowing it to do so without hindrance'. Near Happisburgh, in the east of England, coastal erosion is worsening as a result of rising sea levels and wave action. Buildings close to the coastline are at risk; some have toppled into the sea – twenty-six houses have been lost since the mid-1990s. The area has been without sea defences since 1991, when the groynes (structures built at right-angles from the coast that catch and trap sediment in the surf zone near the shore) and revetments (structures designed to absorb wave energy) below the cliffs were partly smashed in a storm; the rest were removed. The local community has asked the government to build sea defences but the

REDUCING THE EFFECTS OF CLIMATE CHANGE IN TUVALU

Tuvalu, an island group in the Pacific, is one of the lowest-lying countries in the world. During 'King Tides', the land becomes almost level with the ocean; waves break across the islands and salt water seeps up through the soil. One of the Tuvaluan islands disappeared completely in the late 1990s. The Tuvaluan government has been campaigning about rising sea levels for years, firmly linking them to climate change.

Many Tuvaluan farmers now grow their staple crop, a root called taro, in tins filled with compost, rather than in the traditional pits, as salt water has damaged the soil and causes the plants to rot. Taro is also being replaced by more salt-water tolerant plants. The islanders are economising on water, to adapt to drought, and have started to build their houses on stilts, against flooding. Tuvaluans are even reducing their (minute) emissions of greenhouse gases, in the hope of shaming greater polluters into following suit.

The potential for adaptation is limited on an island just 26km^2 in area and whose government has little money for infrastructure development. Sea walls and sand dredging could prevent flooding but dredging alone would cost about £1.3 million.

Official policy is to assist those who wish to emigrate but to continue to work for Tuvalu's future. However, four thousand people have already emigrated to New Zealand and 10,500 more plan to follow. The government is negotiating migration rights to New Zealand for up to half of the country's population in the event of serious changes.

government has declined to do so. Neither, so far, has it paid any compensation, claiming that to do so would create a perverse speculative market in coastal properties.

Other governments have reduced the adaptive capacity of their people by not developing and enforcing appropriate legislation to manage risk. In 1992, Hurricane Andrew hit Florida and exposed the poor building practices of southern Florida: an estimated 40% of the US$16 billion worth of damage was due to shoddy construction. In response, officials improved the building code and south Florida became a model for coastal construction as far away as the Caribbean and Australia. However, in an effort to save homeowners money, in 1999, officials decided that south Florida would adopt a new, weaker, state-wide building code, that required less reinforcing steel in construction, no mandatory hurricane shutters and a weakening of requirements designed to guard against high winds, undoing what was accomplished in the wake of Hurricane Andrew.

Sensitivity and adaptation

Beyond government policy, individuals and communities will vary in their sensitivity to climate change's impacts. Sensitivity to climate change is determined by many of the same factors that contribute to adaptability. People with low incomes are most vulnerable: poor access to resources, inadequately-developed markets, high transport costs and hence reliance on external economies and small domestic markets make individuals and their communities susceptible to external shock and sensitive to the effects of climate change.

Consider two coastal dwellers, both affected by a particularly strong storm that blows coconuts off the trees and batters their houses with high winds. One lives in a house made of concrete, with reinforced steel foundations and roof braces, sited well

CUYAHOGA COMMUNITY COLLEGE
EASTERN CAMPUS LIBRARY

Figure 5 Conditions leading to fewer impacts of climatic change

away from trees and on the slopes of a hill; the other lives in a house made of corrugated iron, lacking strength against the wind, directly under a tree and very close to the sea. Who will be more vulnerable to storms?

The capacity to adapt to change is not the prerogative of the rich and well-resourced and those living in stable environments, although those who have greater access to economic and social resources and information, and the capacity to use them, are more likely to be able to cope with shocks and hazards than those who do not. But people living in areas exposed to high risk, and highly sensitive to change, can still build their capacity to cope with the impacts of hazards. If nothing else, those with few resources have no choice but to adapt.

People's ability to cope is reduced by, among other things, poverty, rapid population growth, migration, inequities in land ownership, unequal access to education and reliance on subsistence agriculture in marginal lands. Unsafe location of buildings and communities, bad housing, malnutrition, unemployment, under-employment and illiteracy all limit resilience. Ben Wisner of Oberlin College, Ohio, and his colleagues have noted that very localised extreme weather, if intense, can cause as much devastation as widespread effects, among unresilent communities. An intense February rain and hail storm in La Paz, Bolivia, which lasted just 45 minutes, led to a surge of floodwater up to two metres high through the streets of the city. The flash floods killed 63 people and injured 146; 5,000 people lost their homes.

Robust local networks can compensate for a lack of resources and help people overcome the worst impacts of weather, irrespective of their level of exposure and sensitivity. Recent research in Cuba and the Cayman Islands has shown that small islands affected by very powerful and very damaging hurricanes recover more quickly when strong local response and recovery groups are involved. It does not seem to matter if the networks are directed by the government (as in Cuba) or collectively organised by citizens, government and the private sector (as in the Cayman Islands).

The Cuban preparedness for hurricanes depends more on education, training and social relationships than on costly procedures and resources, with priority given to people's safety, over and above economic development. In 1976, civil defence training became mandatory for all adults and since then, the government has worked on building its people's skills and organisational abilities. Every year, everyone undertakes a two-day training exercise to test the government's 'five-phase' system. This reinforces peoples' knowledge of their roles and gives them the opportunity to practice any modifications the government may have made to emergency plans in the light of past events. It also allows time to be spent planning and preparing; checking the safety of structures, trees and so on.

1. Information Phase: the media keep the population informed and the government activates its organisational structures.
2. Alert Phase: a full mobilisation is called; students are sent

home, vulnerable people moved to the safest houses or community buildings in their neighbourhood and crops are harvested, if there is time.
3. Alarm Phase: the population goes to shelters and stays there until the hazard passes.
4. Recovery Phase: clean-up teams go into action, with priority given to ensuring the supply of safe drinking water, and civil defence parties assess buildings' structural integrity.
5. Evaluation Phase: a census of the damage is compiled.

Without strong local networks, recovery from disaster can be a longer and more painful process, as seen in New Orleans after Hurricane Katrina. Other US states, such as Florida, had already learned, through painful experience (for example following Hurricane Andrew in 1992) that encouraging people to take personal responsibility for their hurricane preparedness is important for recovery. People are encouraged to understand how their homes are vulnerable to the different hazards of hurricanes (winds, flooding, flying debris), ensure their home is as secure as possible, identify the safest parts of their home, determine their best escape route, including long-distance evacuation, nominate one person within the family as chief contact and know how to get hold of the National Oceanographic and Atmospheric Administration's detailed weather reports. Whichever model of preparedness is adopted – government driven, community-driven or individual-driven – the key is making a plan and being prepared to take action.

Climate change will not necessarily lead to disaster. Disasters unfold when hazards, social and environmental shocks, and economic and political pressures combine to affect people who are in any case vulnerable or unprepared (see Figure 6). Making preparations for managing risk and hazard and reducing people's vulnerability enables them to lessen the adverse effects of climate change and to increase their potential to cope.

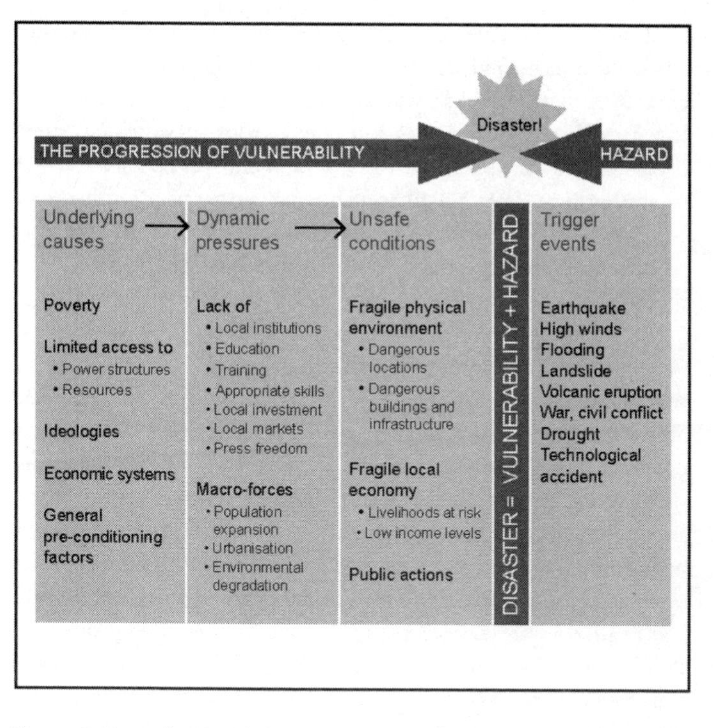

Figure 6 How disasters happen (source: adapted from Blaikie et al., 1994)

Economic growth and development

Climate change is likely to affect national economies in a variety of ways, for example through changes in the market prices of goods and services and in the ability of people to provide those goods and services.

Changes in rainfall, runoff and temperature will affect yields, annual output and varieties of food crops, as well their prices. For example (especially in the southern hemisphere) there is likely to be less water available for irrigation: advance planning could help farmers cope more easily. However, many factors could limit farmers' ability and willingness to adapt by switching to crops that require less water: such crops may have a lower market value, or be less commonly used and hence harder to sell or the farmers may be unfamiliar with how to grow the crop and maximise their yields. As fish species migrate north, fisheries will change; benefits to the north Atlantic fishing industry could be cancelled out by disadvantages to its southern counterpart.

As summer temperatures soar in southern parts of Europe, summer tourism may shift north. As mountain snow disappears, winter sports may become less viable; ski resorts and areas drawing most of their income from this will have to find new economic opportunities. Warmer winters are likely to reduce winter heating requirements but higher summer temperatures are likely to create greater demand for cooling. Energy suppliers need to be prepared and make certain that they can continue to meet peak electricity consumption needs to ensure that industry and businesses are not affected.

Climate change also affects the productive capacity of people and capital (machinery and land). None of these effects are trivial: people's health, and therefore productivity, the availability of energy for industrial production, the productivity of land will all feel the stress of climate change. The changing incidence of disease and patterns of infection will have significant effects: for example, the spread of HIV/AIDS in Africa and Asia has shown how disease can decimate the workforce and slow economic growth and development.

Nonetheless, despite the potential threat of climate change to businesses and economic development, very few businesses account for it in their planning. Many businesses view climate change as a long-term phenomenon, whose impacts will be felt 60 to 80 years hence, rather than being of immediate concern. With the exception of infrastructure building and mineral exploration and development businesses, almost no companies plan more than ten years ahead. Indeed, the majority of businesses consider the medium term to be three to five years.

Conflict and security

Climate change clearly has the potential to create or exacerbate issues of security and conflict. Environmental conflicts, in which companies and countries compete for the rights to extract minerals, oil, timber, fish, water and other natural resources are already common.

India and Pakistan spent ten years trying to resolve the rights for water extraction from the Indus river. After Partition in 1947, tensions flared, as India controlled the flow of the river into Pakistan. The dispute was finally resolved in 1960, when a very costly (US$ 893.5 million) intervention by the World Bank led to the signing of the Indus Water Treaty and the creation of the Indus Basin Development Fund Agreement. Similar conflicts over existing resource use can be seen in other border areas. Since 1944, the International Boundary and Water Commission for the United States and Mexico has been trying to find ways to manage the conflicts associated with water flows to the US and Mexico from the Colorado River, which forms part of the US–Mexico border. In the Mekong sub-region of south-east Asia, the economic slowdown of the late 1990s exacerbated conflicts and disputes over non-renewable natural resources. Desperate to turn their economies around, several south-east Asian governments sought to acquire control over the natural resources of neighbouring countries. In some cases, competition for natural resources has resulted in bloodshed: in the Democratic Republic of Congo, rebels use revenues from illegal diamond sales to buy arms and fuel conflict.

As climate change affects natural resources, agricultural yields, the availability of fertile land, incomes and livelihoods, the stakes will rise and the potential for conflict will increase. Significant numbers of 'environmental refugees', in search of a more viable life elsewhere, will attempt to move from badly-affected parts of the world. Such migration could provoke instability and end in conflict. These are not dramatic stories from a Hollywood movie but today's realities: realities that may worsen significantly.

4

Managing the causes and consequences

Climate change can seem overwhelming: so many people, undertaking so many activities (such as industry, transportation, forestry, land use change, agriculture), producing so many greenhouse gases and so much of them. How can we control emissions?

Based on historic emissions levels, scientists estimate that even if we were to stop producing all greenhouse gases completely today, climate changes would continue for thirty to forty years. Stopping greenhouse gas emissions today is completely unrealistic: we rely on fossil fuels to power industry, commerce, infrastructure, transport, telecommunications and our homes. It will take time to reduce this dependence; in the near future, emissions will continue to grow and the climate will continue to change. We must consider what emission reduction options are available and identify what we need to do to protect ourselves from the effects of climate change now and in the future.

Mitigation

In the context of climate change, 'mitigation' refers to the reduction of emissions of greenhouse gases and the removal of greenhouse gases from the atmosphere.

There are many different schools of thought regarding mitigation. One of the most radical – and the least favoured by most western countries – argues for tackling climate change by making changes to the global economy. To reduce greenhouse gas emissions, it is claimed, we need to tackle the widespread and growing consumerism that is the engine for sustained economic growth in most of the developed world and a widely-held aspiration in the developing world. Advocates of this approach push for a differently-functioning world economy; with a greater role for states and a significant weakening of the private sector. They argue that emission levels are being driven up by an apparently insatiable desire for more consumer goods at low prices, and hence global capitalism (which exploits the most impoverished people to provide these goods at lowest cost) is the root cause of climate change. The solution, they contend, is for everyone to consume less. If workers were paid fairly for their goods, the cost of goods would rise, people would be less inclined to over-consume, waste would be reduced and resources would be more evenly distributed. And if fewer cheap products are moved around the world to meet the demands of the wealthy, greenhouse gas emissions would be significantly lower.

Another – more conventional – school of thought is that we can continue to live as we do at present but make minor adjustments to the way we live, for example changing power sources or developing technology to enable us to live with a changed climate. However, to do so, we need a better understanding both of how we use and manage energy and electricity and how we manage the climate.

Where do emissions come from?

Not everyone is responsible for the same level of emissions. Per head of population, industrialised nations produce far more greenhouse gases than less-developed and less-industrialised countries. The high-emission, developed nations are referred to

Table 1 Annex 1 countries

Australia	Denmark	Ireland	Netherlands	Slovenia
Austria	Estonia	Italy	New Zealand	Spain
Belarus	Finland	Japan	Norway	Sweden
Belgium	France	Latvia	Poland	Switzerland
Bulgaria	Germany	Liechtenstein	Portugal	Turkey
Canada	Greece	Lithuania	Romania	Ukraine
Croatia	Hungary	Luxembourg	Russian Federation	United Kingdom
Czech Republic	Iceland	Monaco	Slovakia	United States of America

by the IPCC as 'Annex 1' countries (see Table 1). Annex 1 forms part of the United Nations Framework Convention on Climate Change (UNFCCC), to which almost all the countries of the world are signatories (with the exception of Andorra, Brunei Darussalam, the Holy See, Iraq and Somalia). The UNFCCC sets out countries' emission-reduction obligations. Although Annex 1 countries account for only 20% of the world's population, they produce about 57% of global gross domestic product and, at present, about 46% of global greenhouse gas emissions.

Many Annex 1 countries have put greenhouse gas emissions reduction policies in place, for example offsetting carbon dioxide emissions from travel, emissions trading (within the European Union), taxes on fossil fuels, information for home owners on how to reduce their energy bills (and thus their emissions) and incentives for businesses to reduce their energy use. Despite these and other initiatives, emissions continue to grow in these countries. Current estimates suggest that by 2030, their greenhouse gas emissions will be between 25% and 90% more than 2005 levels, depending on the extent to which they are able to make reductions.

Table 2 Top twenty emitters of greenhouse gases in 2004

Rank	Nation	Thousands of tonnes of CO_2
1	United States of America	1,650,020
2	China (mainland)	1,366,554
3	Russian Federation	415,951
4	India	366,301
5	Japan	343,117
6	Germany	220,596
7	Canada	174,401
8	United Kingdom	160,179
9	Republic of Korea	127,007
10	Italy (including San Marino)	122,726
11	Mexico	119,473
12	South Africa	119,203
13	Islamic Republic of Iran	118,259
14	Indonesia	103,170
15	France (including Monaco)	101,927
16	Brazil	90,499
17	Spain	90,145
18	Ukraine	90,020
19	Australia	89,125
20	Saudi Arabia	84,116

There are very large differences in the amount of greenhouse gases produced by different countries. The Oak Ridge National Laboratory in the USA has gathered data on the top emitters of carbon dioxide (see Table 2). It is not necessarily that these countries have the highest per capita emissions. In India and China, for example, per capita emission levels are very low, yet, due to their large populations, their aggregate level of emissions places them close to the top of the league. These figures can be contrasted with those countries which produce the lowest emissions (see Table 3). In the main, these are poor or small countries.

Table 3 Bottom twenty emitters of greenhouse gases in 2004

Rank	Nation	Thousands of tonnes of CO_2
188	St. Vincent & The Grenadines	54
189	Solomon Islands	48
190	Timor-Leste (formerly East Timor)	48
191	Samoa	41
192	Nauru	39
193	Chad	34
194	St. Kitts-Nevis	34
195	Tonga	32
196	Dominica	29
197	Sao Tome & Principe	25
198	Comoros	24
199	Vanuatu	24
200	British Virgin Islands	23
201	Montserrat	17
202	St. Pierre & Miquelon	17
203	Falkland Islands (Malvinas)	12
204	Cook Islands	8
205	Kiribati	8
206	Saint Helena	3
207	Niue	1

Reducing emissions

There are significant differences in the amount of greenhouse gas emission reductions that countries will have to endure if concentrations are to be reduced to a safe level and the worst effects of climate change avoided. How should the world collectively reduce its emissions, who should be responsible for bearing the brunt of emission reductions and who should pay? These debates are unfolding at the international level in the annual UNFCCC Conference of Parties. Many specific recommendations have been put forward, which can broadly be grouped into four options: management of energy demands management of energy supply, carbon sequestration and carbon capture and storage. None of these solutions on its own will be enough to reduce the problems associated with climate change; a combined strategy will probably be needed.

'Carbon offsetting' is seen as a way of reducing or mitigating the effects of emissions. Many voluntary carbon offset companies have sprung up; there are now at least 150 offset providers in the world, although most are in Europe or North America. Through these companies, people purchase offsets to mitigate their personal emissions from travel and other carbon-related consumption at a cost of US$3 to US$30 per tonne of carbon equivalent. The money raised provides financial support for renewable and energy-efficiency projects. This is good news for consumers: offsetting can be supposed to create a good from a bad. However, purchasing carbon offsets may not lead to emission reduction: people who buy a carbon offset may think that they are somehow reducing absolute future carbon dioxide emissions. However, if the project or technology is not new, that is, over and above what already exists, then they are not. If people use carbon offsetting as a way of continuing to travel or use energy in a 'business as usual' manner, they will continue to add to the problem. Many people are not aware of this problem and hence take no additional actions to reduce their emissions.

Management of energy demands

Reducing the demand for energy is critical to reducing emissions. Current models show that energy demand is rising quickly, and will continue to rise in years to come, as rich nations use ever more energy and poor nations increase their energy consumption. The International Energy Authority expects global energy demands to be more than 60% higher than current levels by 2040, with 74% of this growth from non-OECD countries. There are two main ways of reducing energy demand: improved energy efficiency and energy conservation.

Wherever energy is used, there are usually ways of using it more efficiently, stopping 'leaks' from the energy system and not wasting energy. Most energy-efficiency improvements can be made relatively simply, without significant changes in lifestyle. People can continue to live in the same houses, wear the same clothes and take the same holidays but undertake to reduce their energy consumption.

Significant reductions in energy demand can be achieved by more efficient management of the heating (or cooling) and lighting of buildings. Domestically, we can all take simple actions to improve our homes' energy efficiency: using draught excluders to stop warm air leaking out, turning down thermostats by one or two degrees, using energy-saving light bulbs and buying energy-efficient fridges, cookers and freezers. More expensive options to increase energy efficiency include installing double-glazing and insulation. When driving, allowing more space between our vehicle and the car in front means drivers are less likely to use the brakes, thereby increasing fuel efficiency (and saving money). When shopping, we can ensure we buy food and clothes that are fairly traded, to make sure farmers and producers get a fair price. Perhaps this way, we will not view our purchases as highly disposable or buy more than we need. On holiday, we can turn off the lights in our hotel room (in some hotels, the lighting system is linked to the keycard – when the card is removed, all the lights are automatically turned off) and keep our use of towels and linens to a minimum.

Energy conservation is about trying to avoid using energy unnecessarily: unplugging electrical appliances (microwaves, phone chargers, televisions, DVD players and so on), not keeping appliances on standby, or using our cars less frequently.

Each of us could consider how we travel, how we power our homes and how energy is used to produce the goods and services we buy. We could change from a petrol-powered car to a hybrid, battery- or gas-powered vehicle – or even to walking or cycling. In our homes, we can use slightly less carbon-intensive energy sources, such as natural gas; low-carbon energy sources, such as combined heat and power sources; or get our energy from renewable sources, such as hydro-power, wind energy, wave energy, solar power and so on. We could also try to buy more locally-produced goods, avoiding out of season flowers, fruits and vegetables flown in from other countries. Energy conservation requires thought and some behaviour change: cutting down on the number of international flights we take, paying more attention to where our food is grown and our clothes are made to reduce emissions from the transport of goods.

It is sometimes argued that poor countries are too poor to be green (and reduce their emissions) and rich countries are too rich to be green (because they are locked into high energy consumption). Yet rich and poor countries alike can do a great deal to facilitate development through enhanced energy efficiency and the consequent reduced costs of energy consumption.

Managing the supply of energy

The energy company BP estimated that in 2008, fossil fuels accounted for about 88% of global energy use (coal 29.2%, oil 34.8% and natural gas 24.1%), nuclear power 5.5% and hydro-power 6.4%. Emissions can be reduced by switching from high-carbon fuels (such as oil and coal) to low-carbon energy sources (renewable sources such as wind, solar and geothermal power, and those with lower-level emissions, such as nuclear power and natural gas).

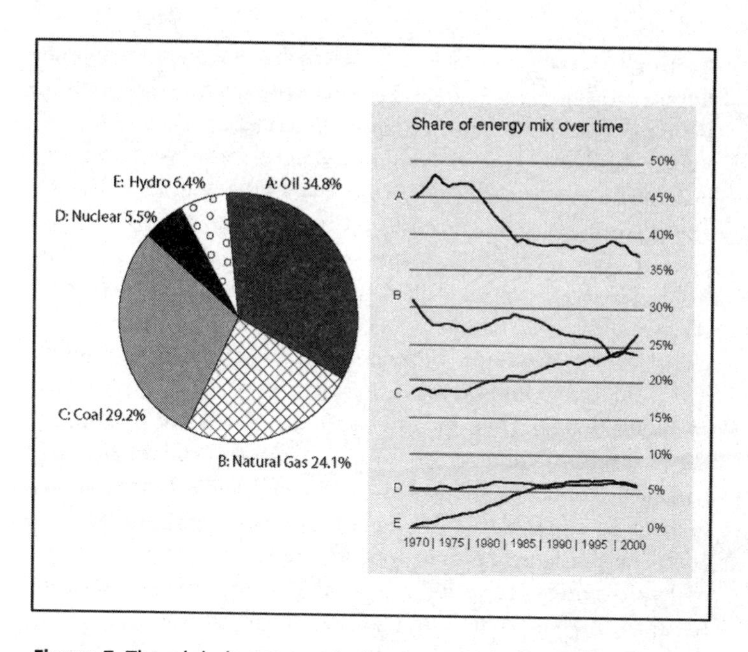

Figure 7 The global energy mix (source: British Petroleum Statistical Review of World Energy 2009, BP plc)

Many countries, especially those that are high emitters of carbon dioxide, rely heavily on fossil fuels. Switching the energy base of an entire nation is complicated; national economies rely on having an infrastructure that supports trade: industrial and commercial buildings, road, air, sea and rail transport, power stations and an energy transmission network. This infrastructure requires continual re-investment. Making the leap from a highly energy-intensive economy to one that is less energy-intensive requires not only a change of energy policy but also radical change in the structure of the economy, for example moving away from manufacturing or industrial processing towards other, less energy-intensive businesses. Investment in research and development of renewable or low-carbon energy sources is an important step. The United Nations Environment Programme estimates that investment in renewable energy rose from US$80 billion in 2005 to US$100 billion in 2006; a trend which is likely to be maintained as oil prices continue to rise and fears that the world is reaching the end of its oil reserves heighten.

The main renewable sources of energy are wind power, hydro-power, solar energy, biofuels and geothermal energy. Renewable energy sources currently meet less than 20% of global energy demands. Biofuels are fuels made from currently living things, recently dead things or from the waste produced by living things. They include ethanol made from low-quality sugar, palm oil and crops such as corn or the physic nut (*Jatropha*). Some commentators claim that *Jatropha* has the ability to produce 2,000 barrels of oil per square mile per year. Biofuels comprise the largest proportion of renewable energy, followed by hydro-power; the other renewable energy sources add up to less than one per cent. The main challenge of using renewable energy is the intermittent nature of the supply, but it has significant potential, particularly in the developing world, where small energy sources can power local development far away from a significant (and costly) national energy grid. The World Energy Assessment, jointly produced by the United Nations Development Programme, the United Nations Department of Economic and Social Affairs and the World Energy Council, noted that Kenya has the world's highest rate of household solar power systems, with more than 80,000 systems in place and annual sales of approximately 20,000 systems. Some commentators argue that nuclear power may play a role in providing low-carbon energy but there is much debate, not only about the extent to which nuclear power reduces emissions but also about the safety issues associated with the storage of nuclear waste.

Carbon sequestration

Carbon sequestration is the long-term storage of carbon dioxide, through biological, physical or chemical processes. Sequestration involves both 'sources' and 'sinks' of carbon dioxide. A 'source' is a process or activity by which carbon dioxide (or other greenhouse gases) is released into the atmosphere, such as driving a car that runs on petrol or burning coal to make electricity in a power station. A 'sink' is a reservoir that absorbs carbon dioxide from another part of the natural cycle.

Changing land use and management of land is seen as the main way in which carbon can be sequestered. Plants are natural sinks of carbon dioxide: when they photosynthesise, they turn carbon dioxide from the atmosphere into carbohydrates (starches and sugars), their source of energy for growth. Of course, when the plants die, their decomposition re-releases the carbon dioxide. Carbon sequestration projects aim either to maintain existing carbon sinks (such as forests) by slowing deforestation and forest degradation, expand existing carbon sinks through forest management or create new carbon sinks by increasing tree and forest cover. Renewable wood-based fuels can also be used as substitutes for fossil fuels.

Management of land use change and forestry has sparked a polemical debate concerning the uncertainties in measurements of carbon dioxide emissions and the limited information on deforestation rates and forest baseline carbon dioxide take-up. Some scientists have argued that our scientific understanding of the carbon cycle (the exchange of carbon among the Earth's plants and animals, soil, water and atmosphere) is strong enough to allow the inclusion of forests in global carbon-trading schemes. But others counter-argue that the scientific uncertainties are still too high.

Carbon sequestration projects have been established in several countries to test ways of reducing carbon dioxide emissions through trade in certified emissions reductions credits. According to international guidelines, these projects should be treated as a subsidiary means of achieving the objectives of the UNFCCC. The projects should be voluntary, permanent, compatible with and supportive of national environmental and developmental priorities and strategies, make a contribution to cost-effective emissions reductions and be implemented effectively. Carbon sequestration projects should also avoid 'leakage': the net loss of the benefits of carbon sequestration by shifting activity from one area set aside for its carbon stocks to another.

In 2007, international governments agreed that their countries should do more to combat tropical deforestation through the incentives provided by the international community, national governments, donors, NGOs and private companies. New schemes for 'reduced emissions from avoided deforestation' will be tested, in which countries such as Indonesia and the Democratic Republic of Congo and communities in the Amazon will receive incentives or compensation for reducing the amount of forest converted to other land uses, such as soya bean cultivation, and the protection of forests from illegal logging.

Carbon capture and storage

A controversial solution to the problem of carbon dioxide in the atmosphere may be to find ways of storing it underground or under the sea in certain rock formations. It would require a significant volume of space to store the vast quantities of carbon dioxide currently being produced and only certain types of rock are suitable. One proposal is to use old, depleted oil and gas fields as storage sites for carbon dioxide. A 1996 study suggested that it may be possible to store 5,300,000,000 tonnes of carbon dioxide in depleted UK oil fields alone. In 2002, 20% of the UK's total carbon dioxide emissions came from sixteen power stations, four steel manufactories and one oil refinery. Where power stations or other large point sources of carbon dioxide are sited above old oil or gas fields, carbon dioxide can be removed from the flue gases and injected into the rock beneath (see Figure 8). There are many challenges to this new technology, not least the risk of leaks from the storage site. If, a hundred years hence, the carbon dioxide leaked out into the atmosphere, this could produce a severe climatic impact for future generations.

A second option is to use saline aquifers (aquifers are underground layers of water-bearing permeable rock); the salty water in saline aquifers is not suitable for consumption or agriculture.

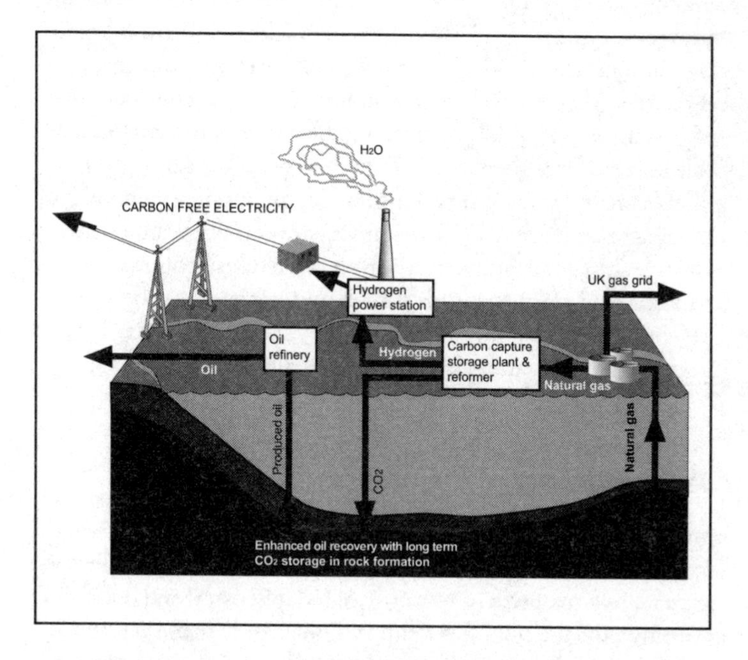

Figure 8 One vision of how a pre-combustion carbon capture and storage plant might work (source: adapted and used with permission from British Petroleum)

It is estimated that, in the UK, approximately 716,000,000,000 tonnes of carbon dioxide could be stored in aquifers. However, little is understood about how the storage of carbon dioxide might affect aquifers and their surrounding rock layers. A third option is to store carbon dioxide in coal seams that are mined out, although even less is known about the risks or viability of this method.

These are all methods designed for capturing carbon dioxide as it is released from large point sources. Imaginative schemes have been devised, such as seeding the oceans with iron filings, to encourage phytoplankton to bloom and so, through photosynthesis, take up carbon dioxide from the atmosphere. However, little is understood about the effects of iron seeding on the oceans or indeed about the ultimate fate of the carbon dioxide. There are strong arguments against such geo-engineering solutions; they are interesting but could have unpredictable and severe consequences.

For many developing countries, exploitation of their coal, oil and gas reserves may be their only economically-feasible energy source. Carbon capture and storage offers these countries a way to mine and burn these fossil fuels and support their economic development without increasing the global concentrations of greenhouse gases. However, this approach comes with risks.

Adaptation

Even if humans were able to wean themselves off their fossil fuel habit instantly, for the next thirty to forty years the effects of climate change will continue to be felt; sea level rises will continue for the next few centuries. We are certain that humanity has to prepare itself for the impacts that are happening now and which will continue to affect present and future generations.

On the surface, adaptation seems straightforward – and far less complicated than mitigation. People will adapt to changes in the weather as they have always done. If it looks like it's going to rain, we take an umbrella and a raincoat, or stay indoors. If it gets hot, we put on our summer clothes. Individual responses to specific weather events are the easy part. The more complicated part is ensuring that our societies are prepared for coming changes and especially, that the least well-off are helped to cope. In such cases, governments must step in with legislation, guidance or a solution that enables the majority to adapt.

Some adaptations, by some individual people, will inevitably affect others adversely. If one householder builds a seawall to protect their property against wave attack, property owners further down the coast could experience worse erosion. And adaptations to one situation may be useless in another. Imagine the following scenario: your summer has been long and hot and the government has put heavy restrictions on your use of water. You develop strategies for dealing with the scarcity of water: you buy a water butt to collect rainwater to irrigate your vegetable plot, use your washing-up water to water the garden and change the plants you grow to more drought-resistant varieties. You put a brick in the cistern to reduce the volume of water used to flush the toilet, take showers instead of baths and replace your power shower with a low-flow showerhead. Then autumn comes, and with it torrential rain. The parched ground cannot absorb the water and there are widespread floods. Your crops are ruined and your home made uninhabitable when the floodwaters mix with sewage from overflowing drains and run through your house. Changing your behaviour to adapt to the dry summer afforded you no protection against the autumn rains.

One of the difficulties in adapting to climate change is that we must prepare for multiple impacts. Some will happen quickly and often: windstorms, heatwaves, tropical cyclones and floods;

others will happen slowly and often, such as droughts; others will be slow and irreversible: sea level rises, rising average temperatures and changing ocean acidity. Coping with multiple hazards makes adaptation difficult to get right. In most cases, making one adaptation (buying an umbrella) will enable you to cope with one type of change (heavier rain) but is unlikely to enable you to adapt to other types of change (summer drought). Somehow, we need to be able to prepare ourselves for all these changing risks.

Disaster risk management

Risk avoidance, risk sharing and disaster preparedness are strategies that enable people to live with changing hazards. Disaster risk management is an umbrella term; an holistic approach that enables people to anticipate disasters and take action to avoid them where possible and, where it is not, to minimise the loss of life, health, property and livelihood. Disaster risk management has four main elements: planning/mitigation, preparedness, response and recovery (see Figure 9).

Mitigation/planning is about long-term risk avoidance; putting in place measures to avoid exposure and making significant risk reductions. In the context of disaster risk management, 'mitigation' takes on a very different meaning to 'climate change mitigation'. In the former sense, mitigation refers to reducing losses through:

- active public education about risk;
- long-term plans to address risk; for example, zoning, land-use management and strengthening building codes and housing regulations;
- tax changes to modify public attitudes to risk; for example, penalties to deter people from building in hazardous areas;
- more effective preventive health care.

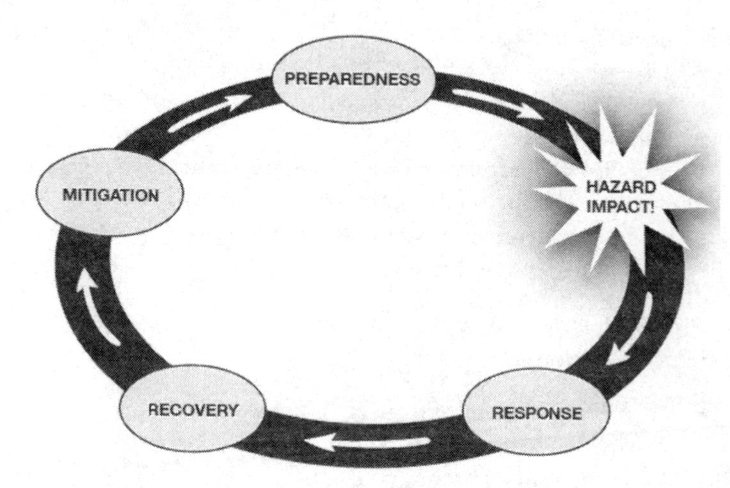

Figure 9 The disaster risk management cycle

Preparedness is very often the only part of disaster risk management that people think about. It is often personal:

- drawing up plans detailing under what circumstances you will evacuate from your house, who will leave, when you will leave and what you will take with you;
- maintaining emergency supplies of bottled water and canned food at home.

Other preparations are part of government responsibilities:

- undertaking emergency exercises and training;
- testing early warning and emergency communications systems;
- public information and education.

Response is the reaction of people or governments to a hazard; what happens when the storm is approaching, the flood waters are rising or the tsunami is bearing down:

- declaring disaster status;
- arranging for evacuation;
- emergency assistance for casualties;
- public warning systems;
- mobilising emergency personnel and equipment, emergency medical assistance and search and rescue teams;
- manning emergency operations centres;
- mobilising security forces.

Recovery follows the event and continues until everything has returned to normal:

- organising payment of damage insurance;
- distributing loans and grants for reconstruction;
- setting up temporary housing;
- providing long-term medical care;
- broadcasting public information;
- imparting health and safety education;
- counselling;
- carrying out economic impact studies.

Together, mitigation, preparedness, response and recovery can provide a comprehensive approach to hazard management and significantly lower the impact of hazards. On 29 April 1991, a Category 4 tropical cyclone hit Bangladesh, bringing a six-metre-high storm surge and winds gusting up to 225km/hour. Despite the cyclone's slow onset (twenty days from genesis to landfall) local early-warning systems were inadequate. Approximately 138,000 people were killed and 10 million made homeless by the storm and the subsequent flooding. In contrast, by the time the equally strong Cyclone Sidr passed through Bangladesh in November 2007, long-term investment in storm shelters and enhanced early-warning systems made by the government, the Red Cross and international donors significantly reduced the people's sensitivity and increased their adaptive capacity. Approximately 3,300 people died, another 1,180 went missing and around 34,500 were injured, yet these numbers, though tragic losses, were but a tiny fraction of what they might have been, had not the Government of Bangladesh made disaster risk management a priority.

Adapting to climate change is not simply about developing risk management plans for every potential hazard. Irrespective of the level of preparedness, there will always be loss. Beyond the costs of preparation, other issues need to be considered, such as how to find the right balance between preparing, for sharing (through insurance or mutual assistance partnerships) and managing risks. In a poor country like Bangladesh, deciding how much to spend on preparations for potentially more – and more intense – storms is a sensitive political issue. Spending more on preparations against disaster automatically means spending less on other equally deserving areas: poverty eradication, health, education, housing and so on.

Not all the risks of climate change will come from rapid and recurrent hazards such as storms and floods. Many effects will start and grow slowly and may not even be noticed until a threshold is crossed. A frequently used, but very accurate, analogy is the story of the two frogs: one dropped into scalding water and one into cold water, slowly heated. The frog dropped into scalding water will instantly jump out, recognising itself to be in danger. The frog in the cold water will stay there even as it gets hotter; it cannot discern the slow changing of the conditions in which is it living until it is too late, its energy is sapped, and it cannot get out.

Slow-onset effects may appear more benign than rapid-onset effects but can prove more deadly. Slow-onset effects of climate change include sea level rises, increased ocean acidity, sea temperature rises, average land surface temperature increases, and falling levels of annual rainfall. From year to year, the changes may not even be noticeable but in the longer term they pose significant threats to homes, livelihoods and life.

Managing slow-onset changes requires long-term planning, focusing both on coping with the changes as they unfold and modifying behaviour as the changes become hazardous. It links risk management with development, raising public awareness, developing early-warning systems, enacting supportive legislation and providing sufficient financial resources. People, communities and governments must assess the potential risks they face and where they are most vulnerable.

Collective action, government direction and international support

The natural and immediate human response to any disaster is to rebuild, as quickly as possible. If your fence blows down in a storm, you are likely to put a new fence back up in the same place; if your coastal home is flooded by a storm surge, you are likely to restore it to the standard it was before. Examples can be found all over the world, from the rebuilding of coastal villages after the Indian Ocean tsunami that struck on 26 December 2004, to the rebuilding of homes and livelihoods in New Orleans after Hurricane Katrina hit in 2005. Such coping mechanisms allow individuals to deal with short-term impacts and continue to function. But while enhancing people's capacity to cope allows them to deal with the hazards they are currently experiencing, it does nothing to change any underlying conditions of poverty, inequality or deprivation. Longer-term adaptations to the effects of climate change need to be considered.

In a desperate situation, few people consider long-term sustainability. If a house is damaged by flood, wind or storm, few would consider relocating; the family needs a roof over its head immediately. In a desperate situation, the link between the immediate effects and climate change might be far from people's minds. And the poorest and least able in society do not even have the choice; they probably don't have the resources to relocate.

Researchers studying how people cope with environmental change point to the need for governments to deliver rapid and effective assistance and for people to work together. Governments are supported by guidance from an increasing body of research (see Figure 10). But the people of a country need to become aware of the risks they face and the reasons they need to act. Public awareness campaigns must be supported and underpinned by good science and clear information and properly targeted. Information that is too broad or too scientific is rarely easily understood.

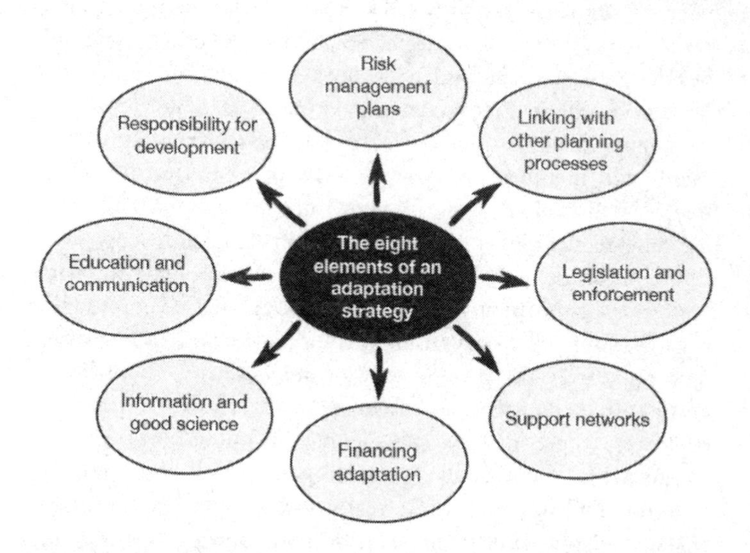

Figure 10 Elements of a government climate change adaptation strategy

Providing information and raising awareness are not enough on their own; someone (or some unit) within government must be responsible for encouraging people to make the necessary adaptations. Without a strong central authority, with the power to reach across departments, it is unlikely that governments will have sufficient resources to deal with adaptation to climate change alongside the many other issues they face. The knowledge that is collated within the unit responsible for adaptation must be disseminated and incorporated into all government processes. This is best achieved through the creation of risk management plans. Governments will need to devise plans to cover all possible weather or climate-related hazards to which their country might be exposed. Yet these plans should not stand in isolation; they must be part of governments' wider economic and medium-term strategic development. Without such integration, planning for the effects of climate change is unlikely to be taken seriously, long-term preparations will be neglected and climate change will only be addressed after disasters have happened and when resources will have to be diverted into ameliorating their effects.

Legislation underpins the behaviour of government departments and members of society. Without clear land use plans, people will build homes in vulnerable places; without good building regulations, buildings will be inadequate to cope with future impacts. Yet even legislation is not enough. In many countries, governments will need to work alongside community organisations, non-governmental organisations, the private sector and even the governments of neighbouring countries, to ensure that disasters are managed effectively. This takes resources, time, money and people. Without governments' commitment, little is likely to happen. Some countries, for example, Finland, France, Germany and the UK, are beginning to make adaptation happen but in many countries, governments do not have the power or resources to implement a strong and co-ordinated adaptation strategy.

And members of society must come together to organise and plan for the possible changes: in the Caribbean, if a hurricane badly affects one country, volunteers from other countries - firefighters, environmental specialists and healthcare professionals travel to assist with recovery. The Caribbean provides some good examples of collective action in response to present-day weather hazards: in Cuba, the Cayman Islands and Grenada, government works hand-in-hand with members of the public, community groups and the private sector to ensure adequate preparation for hurricanes. Poor countries are not totally alone in planning for climate change. Various initiatives, including VARG (the Vulnerability and Adaptation Group, which comprises members of several international development agencies), promote international action and funding to assist developing countries produce national adaptation action programmes.

INTERNATIONAL INITIATIVES FOR SUPPORTING ADAPTATION

The UN Adaptation Policy Framework is a report containing detailed information to assist governments to adapt to climate change. NAPA (National Adaptation Programmes of Action) reports are required under the UNFCCC from Least Developed Countries that identify priority adaptation activities (see http://unfccc.int/ national_reports/napa/items/2719.php). ORCHID is an evaluation tool that enables aid programmes to assess whether they have incorporated adaptation to climate change into their programming. VARG is a group of aid agencies and development NGOs that supports research to determine how best to bring adaptation to climate change into the mainstream of government policy.

Adaptation in the past

The Earth's living systems constantly adapt to the changing economic, social and environmental conditions in which they exist; the climate is just one of the many stresses that affect them. Animals and plants have adapted to the changes in weather and climate many times through the earth's history, in agriculture, forestry, settlements, industry, transport, human health and water resource management both before (some) and after (more) climate impacts were felt.

Although we can be sure that adaptations to climate change have been – and are – occurring, as yet many have not been recorded, systematically analysed or published, because those making the adaptation see no reason to do so. To develop some idea of the range and diversity of adaptations currently being undertaken, researchers are trying to measure and characterise some of the adaptations that have already happened. Adaptations are usually described according to their purpose, their timing in relation to the impact, whether they are short-, medium- or long-term, their size, their effects or outcomes, their form and their performance. The second (1996), third (2001) and fourth (2007) IPCC assessments contain a number of examples of adaptation in practice. Until recently, the link between climate change and adaptation was not necessarily explicitly made: those adapting to floods, droughts, storms and other weather-related impacts may not have made the connection between the short-term event and long-term climate change. For example, the car fleet manager for a large sales corporation may get complaints from sales staff that their cars are unbearably hot in the summer. In consequence – in an adaptation to a warmer climate - the manager might decide to buy air-conditioned vehicles. The higher cost of such cars will be visible in the corporation's accounts but the actual adaptation neither recognised nor specifically stated in the annual report – essentially, hiding it. And many actions are undertaken which provide benefits but are not specifically aimed at adapting to climate change. For example, new buildings may be constructed within improved building regulations, a side-effect of which might be that they can better withstand hotter summers or higher or more frequent floods.

Most governments simply do not have adequate resources to fund their nation's adaptation and many choose to focus on protecting the most needful and those without the capacity to act, and on preparing government itself; its infrastructure, legal and administrative systems, finances and functional capacity. For example, climate change is very likely to increase the incidence of heat-related illnesses. Governments can create early-warning systems, invest in mitigation (for example, in tree-planting programmes to provide shade on city streets) and strengthen building regulations to ensure that new buildings are designed to cope with hot weather or roads made with better materials to ensure they do not melt. They can set up education campaigns about what do to in a heatwave and how to help others, implement vaccination programmes and enforce regulations.

Adaptation is not only about government action. Significant scope remains for people to act for themselves and take responsibility for preparing for the impacts of climate change. People can make sure they drink enough and take work breaks at cooler times of day during heatwaves. Where insect-borne diseases (such as malaria or dengue fever) are likely, homeowners can ensure they don't inadvertently create insect breeding sites by leaving containers of stagnant water lying about, and take preventive action, such as installing window screens to stop the insects getting into their home and buying and using insect repellent and bed-nets.

Since 2000, governments and other agencies have started to track who is adapting, to what and how, with a view to identifying good and bad practice. In 2003, the UK Department for Environment, Food and Rural Affairs collated snapshot descriptions of adaptations to climate change within a six-month period. It is still not known whether the more than three hundred records of adaptation reflect the tip of the iceberg or the entire portfolio of adaptations. The adaptations identified ranged from simple and cheap actions, such as participation in networks, to extensive and expensive engineered defences. It was noticeable in the study that climate change was not a particular 'push' for the adaptations; most were either reactions to weather hazards, responses to regulations and standards such as ISO9001 or examples of corporate social responsibility (especially in the private sector). This suggests that, as recently as 2003, in the UK - a country that has invested significantly more in encouraging its citizens to adapt to climate change than many, most adaptations were driven neither by a realisation that climate change is a problem nor by a realisation that early and anticipatory responses can save lives and keep the costs of adaptation down. In 2008, the US government's Climate Change Science Program commissioned twenty-one synthesis and assessment reports to evaluate the state of adaptation in the USA. Each report identified challenges to adaptation across society, focusing, for example, on climate-sensitive ecosystems, such as marine protected areas, forests, estuaries and wildlife refuges.

The fourth IPCC assessment report divided adaptations into 'proactive' and 'reactive'. Proactive measures included crop and livelihood diversification, early-warning systems for floods, droughts and storms, seasonal climate forecasts, micro-insurance and the expansion of traditional resource management techniques; for example, in Sudan, the importance of traditional rainwater harvesting and water conserving techniques has been recognised. Reactive measures included disaster recovery, relocation of communities or individuals, migration support and emergency response; for example, in Toronto, the Canadian government has set up a 'heat health alert plan' that includes providing designated cooling centres in public places and using the Red Cross to distribute bottled water to vulnerable people. It has not yet been proved whether reactive adaptations will be more costly than proactive; work is being undertaken to find out.

Many adaptations involve building people's capacity to adapt: for example, to help farmers cope with sea level rises and salt water intrusion into farmland, the Bangladeshi government is encouraging the growing of new crops and the use of low technology water filters. In Mexico, in response to drought, farmers are changing planting dates and turning to new crops, such as agave and aloe. In 1997 and 2005, the government of the Netherlands passed a Flooding Defence Act and Coastal Defence Policy that both took precautions against the effects of climate change. The Dutch have also built a storm surge barrier that can cope with a sea level rise of 50cm.

In an ideal world, with perfect information about the timing and distribution of the impacts of climate change, adaptation would be relatively easy. Decision-makers (farmers, politicians, homeowners, firefighters, water managers) could evaluate the costs and benefits of different adaptation strategies to work out which strategy would generate the greatest benefit at least cost, while ensuring that wider social and economic objectives were met. But our world is by no means ideal. Policy-makers operating under uncertainty have to ensure that investment in adaptation is neither wasted nor worsens the way in which the effects of climate change are felt.

There is still some argument about when adaptations to climate change should be made. Lessons can be learned from disaster risk management. The two have many similarities, especially when considering rapid-onset and recurrent hazards. Research indicates that in almost all cases the costs of preparing for a disaster, reducing exposure and sensitivity and increasing capacity to respond are far less than the costs of emergency response and recovery in an unprepared area. Not only are costs of emergency response likely to be significantly higher but the effects are likely to be unevenly felt. Although natural hazards do not discriminate between rich and poor, when hazards strike or disasters unfold the poorest, and those with the least access to resources, suffer most and for longer. Since the mid-1950s, more than one million deaths have been caused by tropical cyclones and earthquakes, about half a million by floods and tsunamis and more than 20,000 by volcanic eruptions and landslides. Centuries of human experience with natural hazards suggests that adopting a 'wait and see' approach to adapting to the effects of climate change is likely to increase the human as well as the financial costs.

Good practice

Good practice in climate change adaptation is not just about disaster risk management. Based on our current knowledge about adaptations to past weather hazards and to the slow changes that we are currently experiencing, preparing for the many different forms of climate change impacts, building adaptive capacity and including climate change impact projections in decision-making is critical.

Good practice is about preparing, gaining as much knowledge as possible about the risks faced, discovering options and ways to fund them and taking climate change into account when making plans, that might be sensitive to climate impacts. Investing in research and understanding means there is a greater chance that adaptations will be right for the challenges faced. Panicked, inappropriate or short-term responses reduce the longer-term capacity to adapt to ongoing climate changes.

Cuba and the Cayman Islands stand out as exemplars, although they have achieved their success in very different ways. Over the years both these governments have invested significant effort in preparing for the hurricanes that annually tear through the Caribbean and affect the islands severely on average once every twelve years. They have made a priority of early-warning systems, communications, long-term risk reduction (for example increasing 'set-back' – the width of the zone from the shoreline in which development is prohibited), preparedness for the annual hurricanes and the incorporation of risk management into government practice and daily life.

The Cayman Islands government has harnessed the power of the private sector to ensure that the National Hurricane Committee, the body that takes control of disaster management in hurricanes, responds equally to both public and private needs. Working in partnership with the private sector, the government makes a priority of getting the island's society and economy up and running as soon as possible after a hurricane. After a storm has hit, building contractors supply resources, such as diggers and heavy plant equipment. This guarantees equipment for the government and provides the contractors with much-needed work. NGOs, faith groups and quasi-governmental bodies such as the Red Cross are fully integrated into the planning, each contributing to the effort and receiving benefits. In meeting the needs of different groups and engaging full participation, the whole population pulls together after a storm to rebuild as soon as possible. The rapid and successful recovery from the Category 4 Hurricane Ivan that devastated the Cayman Islands in 2004 is testimony to this strategy.

The response in Cuba is very different; there, the many arms of government work collectively in long-term planning to ensure that society is prepared and everyone is involved in hurricane preparation and recovery. Civil defence units, with clearly-defined aims, responsibilities and actions, respond as soon as the early-warning system requires. Cuba is frequently hit by hurricanes, yet remains one of the most effective in the Caribbean at preparing and responding with the fewest fatalities.

The UK government's Climate Impacts Programme (UKCIP), begun in 1995, is one example of the effective use of resources to build adaptive capacity. UKCIP began life as a semi-autonomous organisation, with just seven staff and no resources for a research programme of its own, whose remit was to start British adaptation to climate change impacts. UKCIP's initial focus was to consider how climate change might affect the UK and how adaptation might help it to overcome its impact. UKCIP has produced regional climate impact scenarios (which have proved to be very popular) for the UK, updated every five to seven years. Working in partnership with many organisations, UKCIP evaluates how climate change will affect everything from buildings, water supply, tourism and British gardens to biodiversity. It has achieved significant success in encouraging those with whom it works – including government departments, car manufacturers, the utility industries, the insurance industry and farmers – to apply risk-based approaches to adaptation. Much of its time has been spent identifying resources that can help find answers to the questions of how and when the effects of climate change might be felt and how likely they are to happen. This investment of time and resources has raised the UK's capacity to understand climate change, its potential impacts and the risks of inaction. UKCIP contributed to the creation of the 2008 UK Climate Change Act and has heightened levels of awareness about climate change.

Most anticipatory action on climate change is being taken by those industries that will feel the long-term effects of climate change on the outcomes of decisions taken today, notably those which invest in long-lived infrastructure; water suppliers, flood risk managers, energy suppliers and transport companies. Preparing for the effects of sea level rises is in many ways easier than preparing for other effects, as sea level rises are inevitable, irrespective of how we manage greenhouse gas emissions. In Canada, the Confederation Bridge on Prince

Edward Island was built a metre higher than currently required, to take into account future sea level rises associated with climate change; the Deer Island sewage treatment plant in Boston, USA was built on higher ground specifically to avoid possible erosion associated with higher sea levels. Examples can also be found in coastal highway development in Micronesia, in the Copenhagen Metro and in the construction of the Thames Barrier in London.

Bad practice

For every example of good practice there is another of bad. There are many reasons for poor preparation or for ignoring sound scientific advice about the risks faced. Both governments and individual people have a responsibility to encourage action and ensure that relevant knowledge is used to support sound action. Pursuing interests that actively reduce the capacity of others to adapt – obstructive media campaigns, poor communication of the science or selling goods and services that exacerbate climate change – will always be harmful and, at worst, lethal.

One of the key features of such maladaptation is a failure to heed scientists' warnings. The European Environment Agency report, *Late lessons from early warnings*, summarises examples from the recent past where scientific advice has not been taken into account in decision-making and planning. The report details the history of the use of various substances, including asbestos, polychlorophenols and benzene, of which the risks were known, yet were ignored.

The history of the management of asbestos is a perfect example of the pressures that sometimes lead people to take poor decisions in the face of risk and uncertainty. Ten years after asbestos mining began in Canada, in the late 1800s, a possible link between lung disease and exposure to asbestos was identified. Scientific studies began, to assess whether this was a coincidence or a correlation. As more studies were undertaken, it became increasingly apparent that the link was valid. However, the fire-retardant properties of asbestos were highly valued, asbestos was widely used in construction and the asbestos industry employed many people, all reluctant to hear that it was harmful. For fifty years, the battle to acknowledge the link between asbestos and lung disease raged among the industry, employers, employees, health campaigners, the media, the wider public and scientists. Not until the 1970s did governments finally agree that exposure to asbestos was harmful and should be carefully regulated. Yet in 2009, in the UK, there are still asbestos-associated deaths.

Ignoring severe warnings about risks has contributed to disasters throughout history. A recent example of policy-makers failing to heed warnings was seen in the devastation of New Orleans after Hurricane Katrina in 2005. Scientists had provided numerous warnings about the risks that New Orleans faced from north Atlantic hurricanes; a November 2004 paper in *Natural Hazards Observer* described exactly what would happen in New Orleans if a major hurricane (Category 3 or above) hit the city, predicting – almost exactly one year in advance – the impact of Hurricane Katrina and the subsequent disaster.

Bad practice is often easiest to spot in cities. Heatwaves are already a worse experience in cities than in the countryside, due to the 'urban heat island' effect, a combination of reduced cooling through less evaporation of water from plants, less shading as vegetation is replaced by buildings or tarmac, the absorption of heat by buildings and an increase in heat emitted from machinery and vehicles. Building on green spaces simply makes heat islands larger.

Adaptation in developing countries

While adaptation in the world's richer countries is moving ahead, albeit slowly, adaptation to climate change is often seen as far less important in developing countries, whose people, communities and governments face different pressures and more immediate challenges. Poverty, poor or dangerous living conditions, crime, conflict, severe weather and disenfranchisement are all disabling: living in such conditions means that preparing for the worst impacts of climate change in anticipation of impacts becomes far less possible.

Current estimates suggest that climate change is likely to halt or even reverse human progress and make the achievement of international development goals potentially impossible. In September 2000, Heads of State from around the world met at the UN and established their aims for eliminating the dehumanising effects of extreme poverty. Broadly speaking, the 'Millennium Development Goals' agreed at this meeting were the objectives that leaders hoped would significantly reduce poverty by 2015 (source: www.un.org/millenniumgoals):

1. The eradication of extreme poverty and hunger, by halving the number of people who live on less than a dollar a day
2. The achievement of universal primary education, for both boys and girls
3. The promotion of equality between the sexes, the empowerment of women and the elimination of disparity in education
4. The reduction of infant mortality
5. The reduction of maternal mortality
6. The combating of HIV/AIDS, malaria and other diseases
7. The ensuring of environmental sustainability

A 2004 report, *Up in Smoke*, produced collectively by a number of development and environmental NGOs, argued that climate change would affect the capacity of the poorest to achieve the millennium goals and hinder their progress and development in many other areas. Climate change, they argue, has the capacity to stall and reverse opportunities for the poorest to increase their standard of living and find ways out of poverty. The International Rice Research Institute estimates that for every 1°C rise in night temperature, rice production falls by 10%; potentially a disaster for people whose staple food is rice. Many of those living in poverty rely on rainfall to irrigate their crops. In sub-Saharan Africa, the authors of *Up in Smoke* anticipate that crop yields will fall by 20%. About 70% of this region's people are employed in subsistence agriculture, and approximately 90% of that agriculture is rain-fed. The region has had less and less rainfall since the late 1980s and, while farmers have developed ways of coping, it is unclear how, already operating at the very edge of survival, they will cope with increased climatic variability.

Declining yields from subsistence agriculture, coupled with lack of water, poses severe threats to the health and well-being of those in developing countries. The debilitating effects of hotter weather, hazards associated with lack of clean drinking water, the likely increase in diarrhoea, the possible spread of disease-carrying insects and the increasing exposure of new populations to malaria will increase the number of people in need of medical assistance. This will stretch governments' finances and reduce their ability to invest in economic development initiatives, reducing yet further the chances of helping their people move out of poverty.

Many other problems will indirectly affect the potential for growth and ability of developing countries to achieve the Millennium Development Goals. For example, in the annual Asian monsoon, floodwaters often affect the sub-continent. Women suffer proportionally more: in Bangladesh, women are culturally expected to remain within the family and 'covered', yet in a flood, they can be separated from their family or find themselves less modestly dressed than is acceptable and as a result, risk being violently assaulted. Those living in poverty in the Arctic, on small islands and in heavily-populated, low-lying areas and river deltas (such as southern Bangladesh, the Nile Delta and parts of Eastern China) and those living in coastal areas, from Senegal to Angola, Venezuela to Brazil and much of the coast of Indonesia and Pakistan, are at heightened risk from sea level rises.

It is not, of course, impossible for developing countries to adapt but it will require more thoughtful and very careful decision-making. Examples of good practice can already be found; for example, in Mozambique, where bush fires are becoming more common, the German government's aid agency, and others, is providing fire prevention and management training for government officials and communities. And in areas of southern Africa that have already experienced lower rainfall, farmers have started to grow more drought-resistant crops. NGOs and research institutes, such as the member centres that belong to the Consultative Group on International Agricultural Research, International Institute for Sustainable Development and the Tyndall Centre for Climate Change Research, are working with communities to facilitate such necessary change in farming techniques.

Mitigation; harnessing efforts to reduce greenhouse gas emissions by investing in carbon capture and storage technology, alternative energy sources and nuclear power and *adaptation*; preparing societies for the likely impacts of climate change – both options still have many unanswered questions. What is the best solution: building flood defences, developing drought-resistant crops or pouring resources into technological mechanisms for reducing greenhouse gases? Should countries focus on active mitigation, rather than passive acceptance of the results of climate change? Or should they focus on adaptation, because that will bring much-needed funds to those who are already suffering the consequences of climate change? Or should governments take another look at the basic assumptions about mitigation and adaptation as defined by the United Nations? Ultimately, any decision on these questions will be a political act.

5

Navigating the politics of climate change

'We are embarking on a star trek expedition.'

Yves De Boer, UNFCCC Executive Secretary, 2007

The politics of climate change matter, for they dictate the actions that are delivered. Politics is about perspectives, people's voices and the institutions that determine how power is exercised and decisions made. Climate politics in particular is about the legitimacy of public institutions under the threat of climate change, and the accountability and motivation of other agents, such as the private sector and the media, in their contributions to solutions for climate change. And new actors are coming on to the stage: celebrities, businesspeople, writers and broadcasters, social commentators and former political leaders.

Some say that climate change is akin to war. Denial came first, followed by a maturing realisation that something had to be done. To tackle climate change, experts, politicians and people will need to develop considerable mutual trust to ensure that the right actions are taken and everyone pulls their weight.

Social theorists have much to say about how effective climate policies can be put in place. Social theory has developed an understanding of the importance and complexity of changing social relationships and, through analysis of past trends, insight into the future. For example, since the early 1900s, societies in the West have gradually shifted from industrial economies to 'knowledge economies', reliant on formal education. Therefore, those who are not formally educated have gradually lost out to people well-versed in science and technology. Survival in the face of climate change is likely to demand other shifts in expertise, without excluding many more people from the new 'climate-knowledge' economy.

Within the UN, an option exists for collective agreement on new emission reduction targets, beyond the 2012 targets agreed in Kyoto in 1997. This option includes countries currently in a rapid phase of development, such as China, whose economy grew at a rate of eleven per cent in 2007 and India, whose economy grew at nine per cent. However, despite such a phenomenal growth rate, 700 million Indians still survive on less than one (US) dollar a day. It hardly seems fair that India should have to agree to cut its greenhouse gas emissions in 2009, when its per capita emissions are still very low. An alternative route is to allow open market trading of carbon permits. By capping carbon emissions and auctioning carbon permits, the price of carbon should rise, thus reducing demand for carbon-intense products. However, the concept of the carbon market is vigorously contested; some question the logic of allowing companies to continue to create greenhouse gas emissions, suggesting that a better option would be to encourage heavy emitters to move to other technologies. Such a change could either be mandatory or supported through subsidies and tax concessions. At the UN Conference of the Parties (in Copenhagen in 2009), the wealthy nations set out to agree on new reduction targets, as the current period within which they have to reduce emissions ends in 2012. While Copenhagen failed to achieve this goal, countries agreed to stabilise the global temperature at 2°C. The rich countries also agreed to support a climate fund of US$100 billion per year from 2020 to help developing countries tackle climate change, and some rich countries supported a fund for forest protection. Other items that remained unresolved included reforms to the global carbon market and agreement on emissions targets.

Global climate politics in context

Some environmentalists blame the UN for failing to deliver effective action on climate change but the international stalemate is perhaps more a reflection of how international governments function. The founding principles of the UN (created in the 1940s, following World War II) are security, trade and good economic governance. 'New diplomacy', a term coined by Sweden's former environment ambassador, Bo Kjellen, captures the new and uncertain threats posed to societies by climate change, desertification and the loss of biodiversity. Kjellen argues that international institutions are essential in establishing the right conditions for democracy but limited in their ability to bring about action. The revival of Agenda 21, a well-supported 1992 United Nations initiative for managing human interaction with the environment at every level, from global to local, is perhaps a necessary step towards effective action.

National governments collectively agree the UN's rules and are responsible for interpreting, putting into effect and making sense of the broader UN mandate. But most nations are much older than the UN and many national systems are deeply saturated with Western democratic norms of fair and free elections and freedom of speech. National governments are influenced by the politics of those in power and their positions on trade, foreign policy and social and environmental management. National government officials, within their different departments and sectors, have limited resources to undertake national tasks and even more limited (if any) resources to implement UN initiatives. Ministries with the responsibility for implementing international policy – such as environment or rural affairs – are often lowly-placed within government hierarchies. They must compete for resources with health, education and security ministries and find themselves unable to obtain the funds to deliver climate change policies. This is perhaps

CUYAHOGA COMMUNITY COLLEGE
EASTERN CAMPUS LIBRARY

understandable; most societies are built on principles of social and economic welfare. Consequently, a government can be supportive of UN initiatives at the international level but unable to implement them nationally.

The rest of society (NGOs, the media, private companies, community groups and consumers) fills the national political space with contentious struggles over the values and meaning of climate politics. It is in the nature of democracy that topical, highly-visible issues sometimes receive more attention than important but unrecognised risks; this is how climate change issues are sidelined by debates over economic growth. We see three key political challenges: how best to harness the UN to tackle the growing problem of climate change; how governments can best assist their citizens to adapt to climate change, as well as deal with other pressing economic and social issues and stimulate sustainable technology development; how the non-governmental sector can best support or contest international and national policies; what celebrities can do to support adaptation efforts in developing countries; and how the media can be encouraged to cover climate change more broadly.

UN climate politics

One hundred and ninety-four countries take part in the UN climate negotiations, organised in several negotiation groups. The G77/China group, which embraces 130 developing countries, was formally established at the 1964 UN Conference on Trade and Development by the (then 77) independent developing countries. Within the G77/China block are further subgroups: the Africa group, comprising 53 members; the Group of Latin American Countries (GRULAC) with 33; and the Alliance of Small Island States (AOSIS) with 42. The Africa group and AOSIS primarily concern themselves with climate change issues, while GRULAC makes a priority of economic development opportunities. In 1997, in Kyoto, an informal group consisting of sixteen Latin American countries including Bolivia, Chile, Costa Rica and Colombia, was created, which had as one of its main objectives the inclusion of carbon sinks in policy, to ensure the benefits derived from the use of forests as carbon sinks. This group was disbanded in 2001 after the resumed session of the sixth UNFCCC Conference of the Parties, held in Bonn, Germany.

Sub-groupings also exist within the developed, Annex I, countries block (see Table 1). The EU, a regional organisation of 27 countries, with far-reaching economic integration, is the biggest unified political block with regard to population and economy in the developed world. The EU has a relatively global outlook towards environmental responsibility and, following its own experience of acid rain and ozone depletion in the 1980s, follows a 'precautionary principle' in its environmental policy. The EU has established a solid long-term climate policy and maintains a constant leadership in the UNFCCC. The 'Umbrella group' is a loose coalition of countries, with somewhat divergent views on climate change, covering the United States, Japan, Iceland, Canada, Australia, Norway, New Zealand, the Russian Federation and Ukraine.

Due to its size, political power and emissions (almost twenty-five per cent of the world total – see Table 2) the US plays a central and – unfortunately in recent years – negative role in global climate negotiations. Japan also plays a pivotal role, as a high carbon dioxide emitter. Recent US negotiators, framing climate change in domestic terms, have favoured a cautious approach, justified by the highlighting of scientific uncertainties and the costs of emission reductions.

The political debate

Between 1972 and 2008, international discussions on climate change sought to identify the problem, how to fix it and who should fix it. The *problem* was quickly whittled down to the issue of the release of carbon dioxide into the atmosphere from burning fossil fuels, leading to a rise in mean global temperature. The *fix* focused on stabilising atmospheric concentrations of carbon dioxide at 450ppm, beyond which worsening climate change was expected. Focusing on 450ppm also defined *who* should fix it – setting greenhouse gas emission targets for individual countries.

Ironically, given the US government's climate policy under George W. Bush between 2001 and 2008, in the 1980s, American researchers were the driving force in climate change science. The World Climate Conference, held in Geneva in 1979, began a series of international scientific meetings and events leading up to the diplomatic negotiations for a Climate Convention in 1991. In 1988, the Intergovernmental Panel on Climate Change was created by the World Meteorological Organization and the UN Environment Program (UNEP). Under the guidance of its Executive Secretary, Mustafa Tolba, UNEP also launched the UN biodiversity negotiations.

The idea of a multilateral international climate convention was first raised by Malta in the UN General Assembly. In 1990, the General Assembly proposed a Convention and diplomatic negotiations started in 1991. In September 1991, the first IPCC assessment report was published, followed in December by the Second World Climate Conference in Geneva. Diplomats started preparing in earnest for the United Nations Conference on Environment and Development, held in Rio in 1992, at which the Framework Convention on Climate Change was agreed. The Convention, which came into force on 21 March 1994, created an overall framework for intergovernmental efforts to tackle climate change. It also mentions the importance of adaptation to climate change and the need for developed countries to support developing nations in preparing for change. However, progress on adaptation was slow and largely unsuccessful until 2002, when governments agreed to new funding.

In 1996, the US President, Bill Clinton, and his Vice-President, Al Gore, adopted a positive stand towards the Convention and supported the idea of a legally-binding protocol. However, in 1997, the US Congress voted unanimously against binding commitments – to the joy of the oil industry lobbyists, always opposed to IPCC consensus and instrumental in lobbying against any policy of reducing emissions. Nevertheless, the US accepted the Kyoto Protocol – and Bill Clinton signed it – but the subsequent administration decided not to ratify it. Bush's policies were a tough blow to international climate negotiations. Although many were aware that the hard line taken by Bush and his advisors would challenge the negotiations, they were still surprised by the brutality with which they dismissed the Protocol, while signally failing to provide an achievable alternative framework. Condoleezza Rice, when US National Security Adviser, was noted to have declared the Kyoto Protocol 'dead on arrival'. The American stance sparked a dramatic and rapid response from the EU Prime Ministers and Presidents, who agreed to defend the Kyoto Protocol at all costs. For them, Kyoto was the 'only show in town' and abandoning it would be to the severe detriment of the Convention.

Before the Kyoto Protocol could be enforced, fifty-five parties to the Convention (plus Annex 1 countries representing at least 55% of total carbon dioxide emissions) had to ratify it. Without US ratification, Russia became a key player. Russia remained highly ambivalent for a long time but finally ratified it on 16 February 2005, allowing the Protocol to come into force. Without a ratified Protocol, serious concerns existed regarding

THE CLEAN DEVELOPMENT MECHANISM

The Clean Development Mechanism (CDM) is expected to reduce greenhouse gas emissions by the equivalent of up to 2.9 gigatonnes of carbon dioxide by 2012. It transfers large sums of money between private agencies in the northern hemisphere and carbon projects in the southern, to reward individuals or communities for undertaking activities that reduce carbon dioxide levels in the atmosphere, such as planting trees.

Carbon dioxide emissions are relatively straightforward to measure, quantify and exchange: one tonne of carbon dioxide equivalent in the atmosphere equals one Certified Emission Reduction (CER or CDM credit). By September 2009, 4,200 CDM projects were in the pipeline of which 1,814 had been registered. CDM projects include energy, forestation and reforestation and industrial fossil fuel-switching activities.

The CDM has been associated with a number of controversies, including its effectiveness in realising its emission reduction goals, underscoring the lack of consensus on the degree to which these projects are additional to what would have been happening anyway. For example, CDM projects in India have been criticised for failing to contribute in any additional way to emission reductions. Tension between market efficiency and non-market human development work can also make CDM projects controversial. Some suggest that the CDM is time-consuming, risky and expensive, limiting local development benefits and small-scale projects. The CDM also represents an ideological split between those who see climate change primarily as a problem of consumption and those who consider industry has a responsibility for tackling domestic emissions.

Environmental groups argue that the CDM is a loophole for rich countries to avoid emission reductions plus it allows nuclear power an entry point. Other NGOs criticise the ethical, structural and scientific inadequacies of the CDM, suggesting that while industry benefits from CDM projects, poverty and underlying structural inequalities are not addressed and projects are a way for rich societies and industries to elude their domestic responsibilities for tackling climate change.

the future of the carbon market. For ten years, international negotiations focused on the creation of an elaborate global carbon market for emission trading. The stated purpose of this market is to allow carbon-intensive companies to trade for permits giving them the right to emit carbon dioxide, while capping the total level of emissions produced. It was assumed that those companies which introduced new low-carbon technology more quickly would be able to sell on their permits to companies that were more carbon-intensive or slower in the transition to low-carbon technology.

The existence and functioning of the global carbon market have underpinned three important mechanisms designed to enable economic growth while reducing carbon dioxide emissions. The market mechanisms (also known as flexible mechanisms) are the Clean Development Mechanism (CDM), the European Union Emissions Trading Scheme (EUETS) and Joint Implementation (JI). The EUETS is the largest emissions trading scheme in the world. Joint Implementation allows Annex 1 countries to offset their emissions reductions in other Annex 1 countries.

Neglecting adaptation

In the time since the creation of the UNFCCC, adaptation has never received much attention, despite the concerns of developing countries about the impacts of climate change. Early negotiations stuck on questions such as whether climate change was human-caused or a natural phenomenon, who was historically responsible for it and who should pay for adaptation projects and whether the payment should be made in the form of loans or compensation. The early sentiment was that stabilising greenhouse gas emissions was both a priority and more politically acceptable, but oil-producing countries, such as Saudi Arabia, hijacked the debates with concerns about the effects that reducing fossil fuel dependency would have on their countries' economies. For oil-producing countries, the question of adaptation policies and measures remain tricky issues. However, the IPCC scientific report of 2001, which made it clear that the effects of climate change were already being felt in many parts of the world, sparked new interest among negotiators. Quite quickly, negotiators realised that these effects were felt particularly strongly in developing countries, yet their concerns were kept off centre-stage for a long time.

The USA used adaptation arguments as a way to justify not taking action on mitigation. Its view could be summarised as: 'we can always adapt to climate change because we have the resources to do so'. The prevailing view, especially in developed countries such as the United States, was that voluntary adaptation would happen inevitably, as people and communities used their resources to respond to change. This argument proved unrealistic after Hurricane Katrina hit several southern US states in 2005. The devastating effects of Hurricane Katrina helped to change the public's perception of how – even in rich societies – certain communities can be vulnerable to extreme weather.

In 2002, some progress was made on establishing funds for adaptation, accompanied by much debate on the best place to house the funds and who should be in charge of channelling the millions of dollars to adaptation projects in developing countries. In 2002, three funds that could be used for adaptation projects were established, two under the UNFCCC and one under the Kyoto Protocol, all managed by the World Bank Global Environment Facility. Since then, political attitudes towards adaptation have begun to change. At the 2006 UN climate negotiations in Nairobi, government representatives agreed to a five-year international plan for adaptation work, including examining the scientific, technical and socio-economic characteristics of the impacts of climate change, people's and societies' vulnerability to climate change and possible adaptation projects. Leading experts called for mandatory contributions to adaptation projects to be part of an 'adaptation package', without which the lack of financial and administrative capability would limit their implementation in the developing world.

The Adaptation Fund, a subsidiary of the UNFCCC and part of the Kyoto Protocol, came into being in 2007. The purpose of the Fund is to finance adaptation projects and programmes and it is funded by a two per cent levy of the Certified Emissions Reductions proceeds of the CDM. Conservatively, the Fund has the potential to generate US$11.5 million within the Kyoto Protocol in the first commitment period (2008-2012). However, by mid-2009, the Fund had still not got going, as its methods of operation were still being debated by its board (sixteen signatories to the UNFCCC, from both developed and developing countries). According to the Organization for Economic Cooperation and Development's 2009 report, global development aid expenditure in 2007 was US$119.8 billion, while climate change funds pledged by bilateral and multilateral donors (and not including the Clean Development Mechanism) to date total less than US$4 billion per annum (according to ClimateFundsUpdate). Of these pledged funds, approximately US$0.3 billion is being dispersed every year. According to economists at the World Bank, the CDM-related payments contribute a far larger amount, about US$7–8 billion per year, in transactions (see Cappor and Ambrosi's *State and Trends of the World's Carbon Markets 2009*).

Adaptation monies can also be drawn from the UNFCCC's Least Developed Countries Fund, Special Climate Change Fund and from the GEF Special Priority on Adaptation. In 2007, promised contributions to these funds amounted to US$180 million, although only US$84 million had actually been collected. It is currently speculated that more than US$10 *billion* will be required to fund the most urgent adaptation projects and programmes. However, even this sum is likely to be pitifully small: consider the actual costs of present adaptation projects. The old Thames Barrier cost GBP £534 million (£1.3 billion at 2001 prices); the cost of its replacement (needed, as the old barrier is now inadequate to deal with the effects of climate change) is expected to be significantly greater. If just one project can cost so much more, the actual funds available through international funding agencies are likely to prove, at best, token.

CORE ADAPTATION FUNDING AVAILABLE UNDER THE GEF, UNFCCC AND KYOTO PROTOCOL

Adaptation Fund: The Fund was established to finance adaptation projects and programmes in developing countries. The Fund will be financed with a share of proceeds from CDM and funds from other sources. The share of proceeds amounts to 2% of CERs issued for a CDM project activity. At a conservative estimate of US$20 per tonne for CERs during the Kyoto Protocol's commitment period (2008–2012), this fund has the potential to generate more than US$11.2 million. To date no funds have been disbursed.

Special Climate Change Fund: exists to finance projects relating to capacity-building, adaptation, technology transfer, climate change mitigation and, for countries highly dependent on income from fossil fuels, economic diversification. It complements other funding mechanisms. By January 2007, US$38.9 million had been granted to projects.

Strategic Priority on Adaptation: supports adaptation projects. By mid-2009 this fund was worth US$50 million and was supporting 22 adaptation projects.

The GEF Trust Fund: a multilateral fund worth US$2.4 billion; it funds both mitigation and adaptation projects. In mid-2009 it was supporting 591 projects.

Most countries have funding available for post-disaster reconstruction and risk reduction from other sources than public expenditure: insurance and disaster pooling, bilateral and multilateral development assistance and foreign direct investment. However, much work needs to be done to enable the poorest countries to adapt and existing international funds are not yet adequate to guarantee that support.

The future of climate politics currently hangs on the Bali Road Map, created at the climate negotiations in Bali in 2007. This lays out a plan for the international climate negotiations expected to follow on from the Kyoto Protocol after 2012. The key debates of the road map concern the transfer of technology and the 'greening' of economies. In this phase, developing countries will need to take a central place: core challenges include the current lack of technology transfer to developing countries, the lack of public sector funding and the lack of private sector interest. The UN must establish a transparent regulatory framework for carbon markets, so as to send a clear signal about price to the private investors who it is hoped will provide the funds for 86% of future clean energy technology projects in the southern hemisphere. The annual Conference of Parties, who meet in Copenhagen in 2009, need to strike an international deal that includes emissions targets for greenhouse gases that are considerably more ambitious than the current ones, set goals for an improved global carbon market and guarantee financing for adaptation in developing countries, particularly Least Developed Countries.

Expert perspectives and dominant voices

The debate about climate change has moved from the question of whether climate change is happening to what we should do

THE STERN REVIEW

The Stern Review on the Economics of Climate Change, led by Nicholas Stern, former Head of the UK Government Economic Service, was set up by the UK government in 2005 and reported in 2006. On 19 July 2005 the Chancellor of the United Kingdom announced that Sir Nick Stern was to lead a major review of the economics of climate change. The aim was to understand the economic challenges that are posed by climate change in the UK and globally and how these challenges could best be addressed.

The Stern Review aimed to be the most comprehensive ever into the economics of climate change. Through stakeholder consultation in the UK and internationally the Review set out to gather evidence on the consequences of climate change for energy demand and emissions for the economic growth prospects in both developed and developing countries over future decades. It set out to assess the socioeconomic and environmental risks and the costs of adaptation in both developed and developing countries. It aimed to examine the costs and benefits of mitigation actions, including the role of carbon offsets in the forestry and land use sectors and the potential costs of future technological development. And finally, it also set out to explore the impacts and effective actions of national and international policies and institutions for reducing net-emissions of greenhouse gases and the possibilities of a sustainable global economy through investment in cleaner technology.

The main conclusion of the Stern Review was that climate change presents a unique challenge to global economics but that the benefits of strong, early action on climate change readily outweigh the costs. The review resulted in a report that was submitted to the Chancellor in the autumn of 2006. The report contained an economic assessment of the medium- and long-term societal shift to a low-carbon global economy and an analysis of the consequences for activity timescales and institutional choices. It also provided an assessment of the different approaches available for adaptation to climate change.

THE STERN REVIEW (*cont.*)

The Stern Team has continued its work to disseminate its analysis and findings and advise other countries and regions. In March 2009, at the beginning of the Obama presidency, it organised a symposium in Washington DC, attended by senators, academics and business leaders, with the aim of getting a global economic perspective on US action.

about it. However, it remains a debate largely dominated by experts. In the UK media, scientists and politicians frequently debate the realities of climate change; we hear their voices discussing whether the UK government should focus on building flood defences, developing drought-resistant crops or pouring resources into the technology and mechanisms for reducing greenhouse gases.

The media tend to portray expert opinions as polarised – yet in reality they are not truly opposed. While one conservationist, Tim Flannery, argues that to focus solely on adaptation is 'tantamount to genocide', other researchers, such as the Dutch scientist Richard Tol, argue that adaptation is an urgent moral obligation and mitigation can distract developing countries from dealing with the immediate consequences of climate change. These opposing perspectives are rare and most researchers recognise the need to reduce emissions significantly and to prepare for the seemingly inevitable impacts of climate change.

Significant time has been lost in defining 'adaptation' and putting a price on global adaptation. The main conclusion of the Stern Review, that early action on climate change is economically justifiable, is used to back arguments for mitigation, including technological solutions such as shifting to renewable energy sources, developing carbon capture and storage and above all, halting forest destruction (which comprises about twenty per cent of global carbon dioxide emissions). Carbon markets are currently being pushed as the only solution that can ensure that rich societies and businesses change their behaviour, yet there are many other options. Any realistic adaptation initiative has to include sensible development objectives and suitable criteria for sustainability. Countries such as Haiti or Somalia are so poor that they will never be able to adapt to climate change without help. International funding needs to be able to be channelled rapidly to countries vulnerable to the impacts of climate change. With the limited options available, taxes on the rich are the most obvious way to get funds to the poor.

Another option is to find ways to use the power and dynamism of the private sector. There are many examples of businesses preparing for climate change, for example producing bioenergy from crops and designing better and cheaper products for rainwater harvesting. Businesses keep their costs as low as possible and so have an incentive to develop innovative goods and services that will enable them and others to reduce emissions and to adapt. Identifying means to support innovation, research and development within the private sector is another way of encouraging sustainable behaviour. While industry representatives attend the annual climate change Conference of Parties negotiations and can lobby, they are largely represented by some of the major multinationals, such as Exxon-Mobile, and trade groups, such as the International Atomic Energy Agency, the International Chamber of Commerce, the International Fertiliser Industry Association and the World LP Gas Industry.

Mitigation v adaptation

The way that science is used and interpreted by politicians remains unsatisfactory. The UN definitions of 'climate change', 'adaptation' and 'dangerous' are problematic, as they are, ultimately, politically driven. Adaptation is something that people do for themselves, as they have always done, because how climate change affects us depends on who we are and where we live. Even our best models are not yet fully representative of the complex and dynamic relationship between people, the planet, the atmosphere and ecology.

Pitching mitigation and adaptation against each other as rivals for government investment is flawed. Such a simple dichotomy assumes that they are *mutually exclusive* (rapid measures for long-term gain pitted against long-term measures for rapid gain); that policy-makers and societies have to *chose* between 'mitigation' in the developed counties and 'adaptation' in developing countries; and that public policy is determined *solely* by cost-benefit analysis and trade-offs, thus putting mitigation ahead of adaptation, because it is cheaper.

Mitigation and adaptation are not mutually exclusive. Climate change is a global problem; one that requires collective global action. It doesn't make any difference from *where* the greenhouse gases are emitted into the atmosphere. Wherever they come from, actions – whether encouraged by technology, incentives or taxes – are needed to reduce the emissions at their source. Many of the constraints posed by climate change have to do with vulnerability, perceived risk and barriers to social mobility and access. Who lives in hazard-prone areas? Who owns houses built on flood plains? Who lives in exposed parts of cities? What access do they have to information, goods and services? Perhaps societies need to think harder and about more kinds of adaptations – to think beyond flood barriers and drought-resistant crops: how can we build better institutions to create alternative futures and what kind of decision-making structures and political systems will we need?

Neither is it appropriate to set up a dichotomy between rich and poor: the effects of climate change are felt by both. The UK and Japan are prime examples of rich countries on vulnerable islands where the impact of coastal erosion and floods is highly problematic. In 2007, thirteen people were killed by flooding in major UK cities and several thousands displaced from their homes. The devastation suffered largely by the poor and marginal in New Orleans following the medium-sized Hurricane Katrina shows how current weather hazards (and climate change) can affect a rich nation that thinks it is prepared but is not. Poor and small countries are both vulnerable to the impacts of climate change. In the Pacific and Caribbean, strong tropical cyclones are affecting farmers in many of its island nations. In 2005, in the Pacific, there were sixteen named storms, seven tropical cyclones and two major cyclones. In the Caribbean, in the same year, there were 28 named storms, fifteen tropical cyclones, of which six developed into major hurricanes. The death toll associated with the 2005 tropical storm season in the Caribbean and the Pacific reached the thousands.

China is rapidly becoming one of the biggest producers of greenhouse gases, yet, per capita, greenhouse gas emissions are very small. As a whole, China has a very low energy intensity (the amount of energy required for every dollar produced by the economy) relative to the developed world, yet remains one of the nations most vulnerable to climate change impact, in particular to drought and floods.

Money to support projects for adaptation to climate change is clearly vital. Ultimately, the responsibility for dealing with historical emissions and the immediate impact of climate change on poor nations will fall to the rich. Some have suggested levying an adaptation tax on the rich, tied in some way to overseas development assistance (in addition to the existing two per cent tariff on the Certified Emissions Reductions traded under the CDM), to provide the funds to vulnerable nations. However, it will also be important for governments to commit themselves to better economic and civil governance so that funding reaches the right people. A Tyndall Centre report, released in 2006, looked at the adaptation activities of six major donors, including the World Bank. The analysis showed that current aid provision for climate change is highly unsatisfactory. To improve matters, the climate change community and the development community must share their experience and work together.

Climate change adaptation has to be addressed quickly, no matter the cost. Governments across Europe are increasingly discussing adaptation. The EU has published a green paper; the Swiss and Austrian governments are considering ways in which the skiing industry can change to accommodate a more Mediterranean climate and in the UK, the Department for Environment, Food and Rural Affairs has been actively working on a national adaptation strategy since 2005.

Competing stories

Politicians, social commentators and activists have established roles in inspiring and creating movement in societies but other groups and agents are coming on stage. The media, celebrities, businesspeople, academics and former political leaders: all have something to say about climate change.

The different stories with which social groups align themselves help us to understand people's view of the world and how they are willing to tackle climate change. The tellers of these stories play an important role in stimulating debate and motivating others: the work of former US Vice-President Al Gore has had a significant impact. Most of these agents (see Figure 11) are concerned with mitigation and do not address adaptation or the consequences of climate change. This is, perhaps, due to the way that the public debate has been conceived nationally and internationally: climate change

Figure 11 Swept towards chaos

means many different things to many different people: political rhetoric for change, economic opportunities for businesses, the moral responsibilities of celebrities, societies' failures, justice, and equity.

Table 4 also illustrates the variety of narratives on the problem of climate change and its possible solutions. These perspectives can be whittled down to how we see the world, who we think is responsible and what the appropriate solutions are: for example, the singer KT Tunstall balances her carbon-intensive professional life by supporting carbon-offsetting projects. Critics of this approach include the social commentator Mark Lynas, who advocates stricter emissions targets to support a more rapid transition to a low-carbon economy, delivered by taxes and government regulation. Even the media acknowledge (albeit somewhat sensationally) that: 'We are facing a climate change crisis; the science, politics and mitigation

Table 4 Summary of broad types, stories and actors' perspectives on climate change

Type	Stories	Actors
Political rhetoric	*Surviving climate change requires a global environmental revolution*	Former political leaders (e.g. Tony Blair, former UK Prime Minister). Blair has said that there is a need for a global environmental revolution. Barack Obama, the President of the USA, appears to be following this line of thought with his Green New Deal.
Economic opportunity	*Climate change is the negative outcome of successful development; we need to avoid crisis by mitigating and see it as an opportunity to motivate business*	Business community got on board with tackling climate change in 2007. US national football league planted 3,000 trees to offset the Super Bowl. Richard Branson's Climate Challenge.
Moral responsibility	*We offset our personal carbon emissions to balance the climate needs with a carbon-intense lifestyle*	Celebrities offset their flights by paying 'carbon offset companies' (e.g. the band Coldplay). Others offset by tree projects, such as singer KT Tunstall, the Sex Pistols (who planted 500 trees in Essex), Ronnie Wood, the Rolling Stones (who plant their own woodland in Scotland). Source: British Council

CUYAHOGA COMMUNITY COLLEGE
EASTERN CAMPUS LIBRARY

Table 4 *(cont.)*

Type	Stories	Actors
Failure of the state	*Climate change 'chaos' is inevitable; the solution is stricter emissions targets to aid rapid transition to a low-carbon society; the UN system is failing to ensure this happens*	Environmental NGOs (e.g. Greenpeace, Friends of the Earth), social commentators (e.g. Mark Lynas http://www.marklynas.org/, George Monbiot)
Justice and fairness	*Climate change is an issue of North-South inequality; the solution is strict emission targets plus mandatory compensation to the vulnerable in the South*	Development activists (e.g. Oxfam, IIED), Oxfam publication *From Poverty to Power* (2008), Huq and Alam (2008) *Climate Change Adaptation in Post-2012 Architecture*
Sensationalist crisis	*We are facing a climate change crisis; the science, politics and mitigation solutions to climate change have wider economic and societal impacts; politicians are failing to address all of these!*	UK mass media Coverage of climate change polarises and dichotomises the debate. The majority of reporting focuses on mitigation and the implications for the economy. Few examples of reports on climate adaptation exist (Boykoff and Roberts, 2007).

solutions to climate change have wider economic and societal impacts; politicians are failing to address all of these!'

The common message is that climate change is *real* and we need to *do something* about it. However, notably, none of these narratives emphasise climate change adaptation. A study by Boykoff and Roberts in 2007 showed that the majority of media coverage on climate change focused on climate change mitigation. They suggested that more research is needed on media coverage of adaptation to climate change in countries such as India, Brazil and China. They also noted an increase in media coverage relating to foreign aid for climate change adaptation after the release of the IPCC report of 2007. In particular, they point to two articles written by Andrew Revkin and published in the *New York Times* in 2007: 'Poorest Nations Will Bear Brunt as World Warms' and 'The Climate Divide: Wealth and Poverty, Drought and Flood. Reports from Four Fronts in the War on Warming'.

Does this indicate an increasing trend in media reporting on adaptation? Given that adaptation is a far less 'attention-grabbing' topic than the controversies of carbon offsetting, nuclear power, biofuels, food security and the financial crisis, it seems highly unlikely.

6

Winners and losers

Are we doing enough to stop the impacts of climate change or is it already too late for some? When is climate change dangerous? For whom is it dangerous? When does it become dangerous? Who wins from climate change? Are wins even possible in the long term? Will some people inevitably suffer consequences from climate change irrespective of any actions taken today?

The impacts of climate change will produce both absolute and relative winners and losers. Understanding who will win and who will lose will enable us to assess who is more or less interested in taking anticipatory action to prevent the worst impacts. Identifying winners and losers is a complicated task, due to the uncertainties associated with climate change. Internationally, incentives – grant funds to assist with adaptation – exist, but such incentives might lead some nations to self-identify, incorrectly, as losers from climate change.

Is it too late for some?

Since the first IPCC report on the state of scientific knowledge about climate change in 1990, awareness has grown of the risks associated with climate change and the potential losses that may arise as a result of inaction on emissions. Reports from the IPCC and the Stern Review provide estimates of where the worst losses will be. The Stern Review states that:

> *The cost of climate change in India and South East Asia could be as high as a 9–13% loss in GDP by 2100 compared with what could*

	WATER	FOOD	HEALTH	LAND	ENVIRONMENT	ABRUPT & LARGE SCALE IMPACTS
1 °C TEMP RISE	Small glaciers in the Andes disappear completely threatening water supplies for 50 million people	Modest increases in cereal yields in temperate regions	At least 300,000 people each year die from climate related diseases (predominantly diarrhoea, malaria and malnutrition) Reduction in winter mortality in higher latitudes (Northern Europe, USA)	Permafrost thawing damages buildings and roads in parts of Canada and Russia	At least 10% of land species facing extinction (according to one estimate) 80% bleaching of coral reefs, including Great Barrier Reef	Atlantic Thermohaline Circulation starts to weaken
2 °C	Potentially 20-30% decrease in water availability in some vulnerable regions, e.g. South Africa and Mediterranean	Sharp decline in crop yield in tropical regions (5-10% in Africa)	40-60 million more people exposed to malaria in Africa	Up to 10 million more people affected by coastal flooding each year	15-40% of species facing extinction (according to one estimate) High risk of extinction of Arctic species, including polar bear and caribou	Potential for Greenland ice sheet to start melting irreversibly accelerating sea level rise and committing world to an eventual 7m sea level rise
3 °C	In Southern Europe, serious droughts occur once every 10 years 1-4 billion more people suffer water shortages, while 1-5 billion gain water, which may increase flood risk	150-550 additional millions at risk of hunger (if carbon fertilisation weak) Agricultural yields in higher latitudes likely to peak	1-3 million more people die from malnutrition (if carbon fertilisation weak)	1-170 million more people affected by coastal flooding each year	20-50% of species facing extinction (based on one estimate) including 25-60% mammals, 30-40% birds & 15-70% butterflies in South Africa Collapse of Amazon rainforest (according to some models)	Rising risk of abrupt changes to atmospheric circulations, e.g. the monsoon Rising risk of collapsing West Antarctic Ice Sheet
4 °C	Potentially 30-50% decrease in water availability in Southern Africa and Mediterranean	Agricultural yields decline by 15-35% in Africa, & entire regions out of production (e.g. parts of Australia)	Up to 80 million more people exposed to malaria in Africa	7-300 million more people affected by coastal flooding each year	Loss of around half Arctic tundra Around half of all the world's nature reserves cannot fulfil objectives	Rising risk of collapsing Atlantic Thermohaline Circulation
5 °C	Possible disappearance of large glaciers in Himalayas, affecting 1/4 of China's population and hundreds of millions in India	Continued increase in ocean acidity seriously disrupting marine ecosystems and possibly fish stocks		Sea level rise threatens small islands, low-lying coastal areas & major cities such as New York, London and Tokyo		
> 5 °C	The latest science suggests that the Earth's average temperature will rise by even more than 5°C or 6°C if emissions continue to grow and positive feedbacks amplify the warming effect of greenhouse gases (e.g. release of carbon dioxide from soils or methane from permafrost). This level of global temperature rise would be equivalent to the amount of warming that occurred between the last ice age and today – and is likely to lead to major disruption and large-scale movement of population. Such "socially contingent" effects could be catastrophic, but are currently very hard to capture with current models as temperatures would be so far outside human experience.					

Figure 12 Illustrative impacts at different degrees of warming (source: adapted from the Stern Review 2006:57).

Note: This figure shows illustrative impacts at different degrees of warming. Some of the uncertainty is captured in the ranges shown, but there will be additional uncertainties about the exact size of impacts. Temperatures represent increases relative to pre-industrial levels. At each temperature, the impacts are expressed for a 1°C band around the central temperature, e.g. 1°C represents the range 0.5–1.5°C etc. Numbers of people affected at different temperatures assume population and GDP scenarios for the 2080s from the IPCC. Figures generally assume adaptation at the level of an individual or firm, but not economy-wide adaptations due to policy intervention.

> *have been achieved in a world without climate change. Up to an additional 145–220 million people could be living on less than $2 a day and there could be an additional 165,000 to 250,000 child deaths per year in South Asia and sub-Saharan Africa by 2100 (due to income losses alone).*

The review went on to identify how the impacts of climate change will affect different groups as the temperature rises (see Figure 12). The impact is described for regions and nations, with little detail at the local level. For example, with a 1°C warming, permafrost thawing is expected to damage buildings and roads in parts of Canada and Russia. With a 2°C warming, 40–60 million more people are likely to be exposed to malaria in Africa. With a 4°C warming, agricultural yields will decline in Africa and some areas (such as Australia) will become non-productive.

Despite these predictions of global climate-driven disaster, identification of the actual distribution of the impacts of climate change will not be that straightforward. Local, regional and national assessments tend to identify different winners and losers, because aggregating results across large areas hides the social strengths and weaknesses of the area and geographic variations. It is difficult to make blanket pronouncements about the impact of climate change on different groups. For example, national level assessments of agriculture in the USA show that, overall, the US is likely to benefit from climate change, yet more detailed regional studies show that the benefits are likely to accrue to the richer countries, most of which are in the temperate zones in the northern hemisphere; the poorer nations, which tend to be in the southern hemisphere, are likely to be adversely affected. Not everyone in poorer states will be a loser. Individuals and communities may adapt their farming to the new climate and could be winners or losers depending on crop prices.

Dangerous to whom and when?

Research by Mike Mastrandrea and Steve Schneider (and later by Suraje Dessai and his colleagues) on the meaning and definition of 'dangerous' climate change highlights the fact that 'danger' is not just a scientifically defined term but also a socially and culturally defined concept. For most of us, something becomes dangerous when some 'threshold' has been crossed, or when we personally experience an impact.

Climate thresholds are also both physically and socially determined (see Table 5). Potential physical threshold-crossings could include the disintegration of the West Antarctic Ice Sheet, which would exacerbate sea level rise and threaten the very existence of many small, low-lying island states. The turning off of the Thermohaline Circulation could have deleterious

Table 5 Social and physical thresholds of dangerous climate change

Danger measured through thresholds in physical vulnerability

1. Large-scale eradication of coral reef systems
2. Disintegration of the West Antarctic Ice Sheet
3. Breakdown of the Thermohaline Circulation
4. Modification of crucial climate system patterns such as the El-Niño Southern Oscillation and the North Atlantic Oscillation
5. Climate change exceeding the rate at which plants and animals can migrate

Danger measured through thresholds in social vulnerability

6. Irrigation demand exceeding 50% of annual seasonal water usage for agriculture in Northern Victoria, Australia
7. Depopulation of sovereign atoll countries
8. Additional millions of people at risk from water shortage, malaria, hunger and coastal flooding
9. Destabilisation of international order by environmental refugees and emergence of conflicts
10. World impacts exceeding a threshold percentage of GDP

consequences for northern Europe, as the warm moist air that permits a relatively mild climate would cease to flow. Social thresholds are determined by how well people can protect themselves. For example, fear of future climate change could lead to the depopulation of small islands, which could push the islands to the edge of sustainability and diminish their viability as sovereign states.

How people perceive the hazards thrown at them also affects their concept of 'danger'. For example, to understand the dangers associated with windstorms, consider two different families in neighbouring homes about to be pounded by 145 mph winds. Family A lives in a house built within the local building code which requires: appropriate straps to attach the roof to the walls, braces in the gable end of the roof to keep the roof supported, the roof is tiled and laid to ensure maximum protection and it has storm-proof windows that can withstand sustained winds of 150 mph (the scientifically defined 'dangerous' speed). Next door lives Family B. Their home is poorly constructed and has not been built to meet the building code standards: winds of more than 95 mph would damage the roof and winds of more than 120 mph would damage the walls. To this family, a storm with winds of 145 mph would be very dangerous.

Both families may be terrified by the storm, have little faith in the construction of their homes and suffer stress during and after the storm. For both, the storm is dangerous whether their home is affected or not. Danger is determined as much by the event as by the conditions in which people live, their personal experiences, their perceptions of their safety, their ability to cope, their values, the information available, and cultural and institutional behaviour.

Nonetheless, some risks appear more dangerous than others. Roger Kasperson, Nick Pidgeon and their colleagues have explored the concept of 'the social amplification of risk'. Their work explores cases in which very small risks are deemed unacceptably large by society and very large risks ignored or not considered dangerous by the majority. Pidgeon and his colleagues conclude that perception of risk is more influenced by heuristics (that is, experience-based means of understanding problems, such as how memorable they were or how you felt when you experienced the risk) and the qualitative characteristics of the risk (for example whether you chose to expose yourself to it and whether there were potentially catastrophic consequences) than by scientific evaluation of the risk.

While the effects of climate change are a major challenge and significant threat to global societies, their risks must be considered relative to all other risks. At present, people in richer countries are more likely to die from a lifestyle-related disease than from the effects of climate change; according to the UK national statistics, cancer and heart disease were the cause of 50% of all deaths in 2003. Different cultures and groups have very different attitudes to different risks, although many bodies (notably the media and organisations in charge of managing the risk) influence whether risks are amplified or attenuated. For example, one of the chapters in *The social amplification of risk* (Pidgeon, Kasperson and Slovic) considers two cases of contamination in India, one the discovery of the bacterium *Yersinia Pestis* in the city of Surat, where the risk was amplified significantly, the other a case of arsenic contamination in West Bengal, where the risks were attenuated. In both cases there were similar levels of impact yet the portrayal and perception of the risks was very different. Risks clearly depend on how we feel about them, whether we trust the agencies involved in their management and whether we feel that anyone should be blamed for it. 'Danger' is only partly related to the science underlying the risk. If someone feels that they are at risk from climate change, they are likely to change their behaviour to reduce the severity of its perceived impacts. But they may also amplify the risks by taking inappropriate actions.

Although it is possible we may amplify or attenuate the risks associated with climate change, there is still a valid argument that the threats need to be kept under careful scrutiny. Unless we know that action is needed, little action is likely to be taken. If we fail to fend off the worst effects of climate change, future impacts are likely to be worse. The psychologists Torsten Grothmann and Anthony Patt have argued that people need to feel that they *can* cope with environmental risks and that their effort is *not* futile or they tend to give up hope and resistance. Abandoning efforts to tackle climate change before they have begun in earnest would be a tragedy.

Is loss inevitable?

Even though we know that climate change will not necessarily be dangerous for everyone and that the way in which climate change is perceived will, to some degree, determine whether it is dangerous or not, some actual losses will be inevitable.

Some aspects of climate change will continue whether or not humans manage to reduce greenhouse gas emissions. The atmosphere is warming, the sea is rising, the planet is changing. We know that some people will die as a result of climate change but we cannot say exactly how many. Very few studies describe the impact of climate change on mortality. The most frequently cited, published by the World Heath Organisation in 2002, estimated that about 150,000 people are killed annually as a result of climate change. This figure is based on estimates of fatalities from malaria, malnutrition, diarrhoea and drowning from floods alone; current estimates indicate the annual global impact of climate change is already above this level.

Estimating the impact of climate change loss is challenging for scientists for several reasons: first, identifying a climate change-induced 'cause' of death is difficult; there are many ways in which the climate (and climate change) may cause a person to die – by drowning in a flood, from drought, through famine, by catching an infectious disease, from skin cancer or by coming into contact with a new disease, spread by a previously unencountered carrier. Second, even where it is possible to identify a climate change-induced cause it is often difficult to distinguish between the effects of climate change and normal climate variability. Third, there may be health improvements in some countries as a result of climate change (for example, fewer cold-related deaths in northern latitudes). The life-years gained have to be included on the positive side of the mortality balance sheet.

Loss of human life from the effects of climate change is inevitable. Arguably, other losses are equally inevitable: there will be some loss of land as the sea level rises, the coldest parts of the planet may no longer be able to sustain their flora and fauna as the temperature increases and if they cannot adapt, there will be no cold places left to which these species can migrate. The inevitability of loss depends on several factors: whether the Earth crosses a climate threshold that throws it into rapid climate change to which living things cannot adapt quickly; whether humans can stabilise greenhouse gas emissions; whether humans can reverse the changes in the atmosphere through an engineering solution and whether living things can adapt to the impacts that will cause great hardship, for example, ocean acidification and ocean warming, sea level rises, droughts and floods.

Irreversible loss is a very plausible scenario, even though we cannot assess exactly how likely it is. Lacking such certainty, the most effective – and the fairest – approach to managing climate change is to reduce greenhouse gas emissions as much as possible while also preparing for the damaging consequences that may still take place.

Scenarios of concern

Irreversible loss caused by the effects of climate change is more likely under certain conditions. These include the failure to reach international agreement on stabilisation targets or greenhouse gas emission levels, not acting quickly enough to reduce emissions or to prepare for the effects of increased emissions and the effects of only some countries acting.

What if ... international agreement is not reached on limiting greenhouse gases?

There is little doubt that climate change will worsen, its effects will worsen, the poorest will suffer and countries may have either to take aggressive action to protect their limited resources or stop the flow of environmental refugees. The UN report *Global Environmental Outlook 03* considered four possible scenarios in a world where greenhouse gases are not limited:

Markets First describes a world that relies increasingly on globalisation and trade liberalisation to bring about growth and development. In this scenario, industrialised nations pursue consumer-driven goals that increase the importance of markets in economic growth.

Policy First describes a world in which strong governments direct the private sector to reach specific social and environmental goals.

Sustainability First centres development on new, more equitable institutions, in a world in which values and development paths are more closely aligned to sustainability.

Security First describes a world of inequality and conflict, brought about by socio-economic and environmental stresses.

This last scenario, *Security First*, is as plausible as any of the others, yet this scenario describes a situation in which: 'The global economy remains stratified and fails to embrace the billions who are economically and politically marginalised ... traditional livelihoods and communities also erode' and 'many of the poor try to migrate to rich countries and rising numbers of them resort to illegal entry. Affluent groups respond with growing xenophobia and oppressive policing of borders'. *Security First* essentially describes the worsening of climatic conditions, under which people from poor nations try to escape unproductive lands, for example from Africa and India. The mass migration into Europe and the clamouring of environmental refugees is a media scenario often highlighted as the big future challenge that we face.

Security First is just one of many possible scenarios: there are many others far worse. Countries with access to carbon-based resources (such as the USA, Venezuela and Saudi Arabia) may force a delay in greenhouse gas emission reductions and increase the price of fuel to compensate their nations for a future without fuel income. As a result, fuel poverty may increase and conflict over unclaimed carbon-rich minerals, for example those in the Arctic, may escalate. In the twenty-first century, wars for access to scarce mineral resources may be fought by the developed nations. However these scenarios unfold, they are unlikely to bring about significant improvements in global well-being.

What if ... we do not act soon enough?

The longer we delay action on climate change, the worse the impacts will be. Research compiled in 2007 by the IPCC assessed the timing and distribution of impacts as the temperature rises (see Figure 13).

We do not know how quickly the climate will warm but we do know it will warm. We also have an idea of what the impacts of this warming might be. Water supply, human health and

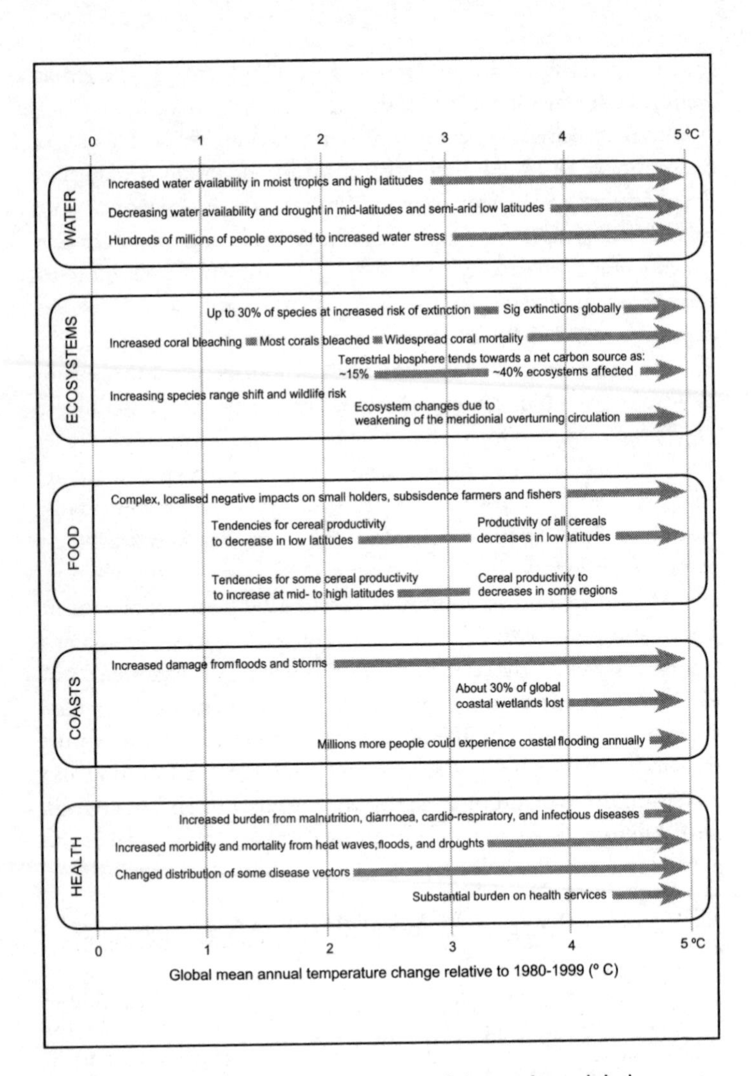

Figure 13 Key impacts as a function of increasing global average temperatures (source: adapted from IPCC AR4, WGII, Summary for Policy Makers, p. 16)

well-being, the health of ecosystems, food chains and food supply, the coast will all be affected.

As the temperature rises, global inequalities will increase. As the moist tropics and high latitudes receive more rainfall and thus an increasing water supply, hundreds of millions of people in the mid- and semi-arid low latitudes are likely to experience water stress and water shortages. Inequalities in human health will become starker. High rates of cold-related deaths will decline in the colder countries; most countries will experience an increasing burden of ill-health due to malnutrition and an increase in diseases spread by animals and pests.

The impact of climate change on ecosystems, food and coasts is not likely to discriminate. All living things will experience food stress as populations compete for limited resources. Countries with the resources to purchase food and the arms to protect food transport corridors will most likely be the winners. The poor, and those even now struggling to maintain their livelihood, will be the losers. Coastal flooding and inundation will affect all coastal countries. As ecosystems struggle to adapt to the changing climate, important biodiversity may be lost, which may reduce the capacity of the environment to support services on which living things rely, such as nutrient cycling and provision of food and timber. If warming reaches 5°C, there will be dangerous consequences for human beings, and most of the ecosystems on which we depend.

What if ... Some countries act but others do not?

One of the main benefits of the international negotiations on climate change is that they bring together all countries to discuss issues that affect all countries. Not everyone has played a part in the creation of the problem but everyone can be part of the solution. In bringing together the perpetrators and the victims of climate change, it was hoped that international agreement could be found. Yet, after fifteen years of annual UNFCCC meetings, there has been neither a demonstrable impact nor any stabilisation of greenhouse gas levels. Should international negotiations continue or should other non-UNFCCC solutions be sought?

Some commentators argue that not all countries need to be involved in the negotiations; that the issue could be resolved by bilateral agreements between the main emitters of greenhouse gases. Small island nations produce negligible emissions: arguably, reductions by these countries will make little difference. If large emitters can reach agreement among themselves about phasing out carbon-based fuels such as petroleum, gas and coal, this could be a way forward.

However, other issues need resolving. First, issues of justice are involved in decision-making about climate change. Some argue that those who caused the problem owe a debt to those who suffer its consequences. Claims are already being made by those who are experiencing the impacts of climate change against those who are causing it: in 2005, a petition was submitted to the Inter-American Commission on Human Rights by the Inuit Circumpolar Conference. The Inuit claimed that, as a result of the American failure to reduce greenhouse gas emissions while knowing the impact of those emissions, the livelihoods, cultures and lives of the Inuit were at stake. In the petition, the Inter-American Commission on Human Rights was asked to investigate the harm caused to Inuit by climate change and, if it were found to be true, declare the US '... in violation of rights affirmed in the 1948 American Declaration of the Rights and Duties of Man and other instruments of international law'. The presence of the vocally active victims of climate change in the negotiations may speed up the process of deliberation to ensure that further debts do not accrue before action is taken.

Second, as Albert Einstein is reported to have said: 'We cannot solve our problems with the same thinking we used when we created them'. The countries that created the institutions, infrastructure, societies and development pathways that rely on carbon-based technology are not perhaps the best placed to develop a new world order or sustainability paradigm. Other minds may be needed to find solutions that take new and creative routes for managing the consequences of the past.

Third, no country's dominance over others remains fixed: their importance and power rises and falls. Historically large greenhouse gas emitters are slowly being overtaken by emerging economies. China and India stand out: two countries, developing quickly and following the fossil-fuel-based growth trajectory of most developed countries. The emerging economies need to be included in negotiations. They are becoming important players on the world stage and their emissions too need to be controlled and reduced. International investment is needed for technology transfer, so that advances in low- and no-carbon technology can be passed on to newly industrialising states.

In short, what is needed is bold action by governments. Internationally, to push for greater action and find agreement; nationally, to develop, implement and enforce climate change legislation in the industrialised world. Such actions would show that there is serious commitment to tackling this problem and provide guidance and leadership to other countries.

How can we cope better in the future?

To cope better in the future, we need to understand the consequences of the different forms of 'governance' of climate change. Governance refers to the process of managing power, authority and influence in shaping decisions. Governance is broader than government, for it includes interactions between government and civil society. The governance of climate change

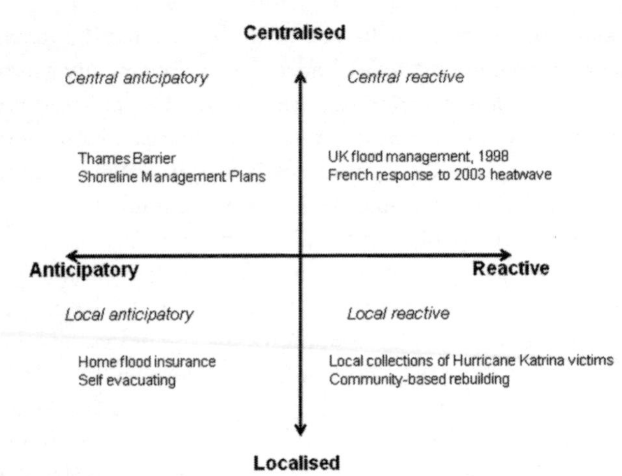

Figure 14 Examples of governance regimes for managing climate and weather hazards
Source: Tompkins, E. L., Brown, K. and Few, R. (2008) Scenario-based stakeholder engagement: A framework for incorporating climate change into coastal decision making, *Journal of Environmental Management*, 88 1580–1592.

relates not just to who should participate in action but also to when action should be taken.

For effective governance, research has shown that a variety of issues need to be considered, including attitudes to risk and uncertainty, the role of government, the resources available and the political and cultural context. The broad spectrum of decision-making styles involved in governance can be simply characterised in four ways (see Figure 14): central anticipatory, central reactive, local anticipatory and local reactive.

Central anticipatory refers to decision-making that focuses on reducing possible exposure to risk by acting sooner rather than later. The risks addressed are identified by central government rather than being locally determined and the costs are borne by the whole population, through taxation. This type of decision-making can be seen in the UK government's approach to coastal planning – the UK Shoreline Management Planning Process. Through this process the government determines which areas of the coast to protect and which to abandon. Another example is the investment in a flood barrier for the river Thames in London, conceived and built following damaging flooding in the east of England in 1953.

Local anticipatory describes governance driven by local needs and actions and organised by local people in anticipation of the impacts of climate change, to which the needs, aims and desires of local communities are central. The risks are locally identified – they may not be the same priorities as the central government's but they are most important to the community. Wherever decision-making power is devolved, funds usually have to be raised locally and so lack of resources can limit this form of governance. Good examples of local anticipatory action can be found in various forms in Bangladesh. One response, by people living on the chars (mobile river islands) of northern Bangladesh, has been the planting of indigenous reeds (*Saccharum spontaneum*) along the banks of the chars to prevent soil erosion and hence reduce damage during floods.

Central reactive describes centralised decision-making taken in response to the effects of climate change. The costs of response can be very high but they are evenly distributed across the population. The French government's response to the 2003 heatwave, which caused thousands of deaths among elderly and vulnerable people, is typical of this response. The government had rapidly to concoct an emergency response to the heatwave, with no plans for guidance. The response was piecemeal; significant criticism was later levelled at the government. A second example is the UK government's response to the Easter floods of April 1998, when very unusual weather led to widespread flooding in the English Midlands. Just five people died, but 4,500 homes, 522 industrial premises, 2,000 caravans and numerous cars were damaged. The total cost of the flood was approximately £300 million. Again, the lack of any plan by the government meant that an ad hoc response had to be developed and implemented during and immediately after the floods.

Local reactive describes local actions taken in response to local needs during an emergency, either as it is occurring or after it has happened. As with 'central reactive', responses occur after the impacts and hence the costs can be very high. Examples can be found in the aftermath of Hurricane Katrina in 2005, when, due to a poor centrally-organised immediate response, collections were made within the region to fund recovery. Community-led rebuilding after hurricanes is typical in parts of the world where resources are limited or central government is weak or disorganised.

What style of governance is adopted will depend on the context in which decisions are taken; none can be considered a universal ideal. Whichever is used, costs will be incurred; irrespective of the type of governance, at the heart of any climate change response is the need to reduce greenhouse gas emissions.

Greenhouse gas emissions reduction strategies: winners and losers

There are many ethical consequences of the mitigation strategies that societies adopt. Any action taken to make a good out of a bad has the potential to make a worse out of a bad as well. People all over the world are asked to reduce their emissions and to find low-carbon ways to live. Unfortunately not all emission-reducing activities are likely to be universally beneficial.

To reduce their 'carbon footprint' (the total greenhouse gas emissions caused by individuals), some countries have chosen to switch energy sources from fossil fuels to biofuels. As the demand for biofuels rises, so does the price of the various commodities used to make them, making them more economically attractive than food crops. The danger is that poor farmers will decide to grow biofuel crops, which may earn higher returns but put them at risk of fluctuations in the market price and the problems of monocropping. Evidence is already starting to emerge that biofuel plantations have destructive consequences for ecological systems and livelihoods. Examples exist across the developing world: the deleterious effects of palm oil plantations in Malaysia, soybean production in the Brazilian Amazon and the conversion of wetlands to ethanol plantations along the coast

FOOD SECURITY AND CLIMATE CHANGE

Food security has been eroding since the 1990s. By 2020, the impacts of climate change will have had profound effects on global food security. Changes in precipitation and temperature rise will reduce the yields of crops such as corn, wheat and rice in semi-arid regions of the world.

Rising food prices and less land area available for agriculture have contributed to weakening food security in poor countries, especially countries where many people rely on crops for subsistence and for extra income from selling their produce. Poor farmers are already vulnerable, because they have limited options in existing agricultural systems. Farmers will need to adapt and change the crops that they grow; for example to drought-resistant crops. However, this may be difficult: farmers might not want, understand how, or have the technology to produce sorghum instead of corn. And their families and customers might not like the taste or know how to prepare or cook the new crop.

There are no easy solutions. The impact of climate change will have profound consequences for societies in poor countries. It may not be enough to prescribe technical and management solutions such as the provision of improved seeds, crops and fertilisers, better land use and decision-making, without better understanding of the relationship between people and their land, crops and food.

of Kenya. While some argue that there is plenty of land available for biofuel crops, others argue that the best land is already occupied by food crops and, moreover, provides a habitat for wildlife. There are many unanswered questions about how to balance the world's climate crisis with its fuel needs, food production and biodiversity protection.

In the short term, climate change will produce winners and losers. It is also likely to reinforce the status quo in power and development, where wealthy countries can use their resources to procure goods that climate change or changed priorities have made increasingly difficult to find. Poor countries are likely to find themselves tied into unsustainable farming practices and trying to appropriate a small share of the profits from the changing demand for crops, or wedded to the world economic system with all the price fluctuations and distortions that can quickly lead marginal farmers into poverty and destitution.

In the long term, there are no winners. New evidence reveals that the changes we can expect during the twenty-first century will not be within the 'safe' margin of a 2°C temperature rise; we could see 4°C of warming or more. Growing climate change and variability, rising sea level, increased competition for food and increasingly limited natural resources make the *Security First* scenario the most likely future to which we are headed. It behoves everyone to consider what we can do to ensure that this doesn't happen on our watch.

7

Forcing change

We believe that there is a strong need for change in many aspects of our society, in our attitudes, our politics and our science, to ensure that human societies are adequately prepared for climate change. Identifying innovative development trajectories, developing appropriate international actions to support collaboration and consensus, implementing national actions to reduce emissions and prepare for impacts and initiating a shift of attitude towards people's personal responsibility for the problem and for dealing with the consequences will all be needed.

Much has been achieved in raising awareness of climate change and convincing both governments and the public that climate change *is* happening, that human behaviour *is* causing it and that to avoid serious consequences urgent action *is* needed. However, so far very little has been achieved in getting governments to cut their emissions and make funding for adaptation available. Disaster risk reduction strategies and action on greenhouse gas mitigation through development of the CDM can help improve the energy efficiency in power grids. Installation of new technologies and the reduction of wasted electrical energy by outdated power delivery and distribution systems also require attention.

Twelve years on from the Kyoto agreement, society is in transition. The national obligations established under the UNFCCC are starting to affect social choice and behaviour on a grand scale. Several hundred businesses have signed up to the Carbon Disclosure Project, which aims to publicise each company's carbon footprint. In the United States businesses and cities have supported the idea of a national 'cap and trade' system and companies are preparing for the day that legislation on climate change is enforced in the US.

The International Energy Agency (IEA), which has 27 member industrialised countries, is at the forefront of thinking about how to cut carbon dioxide emissions by 50% by 2050. Controversially, the IEA has said that 1,400 nuclear power plants and a large number of offshore wind farms will be required to achieve this goal and that funding (possibly in the form of government subsidy) of around US$45 billion will be needed. Other research institutes, such as the Tyndall Centre, have suggested that the target can be met without becoming reliant on nuclear energy, through better managing energy demand, regulating high-emission transport and developing alternative renewable energy sources.

Important influences on public attitudes and government policy on climate change include Al Gore's 2006 documentary *An Inconvenient Truth*, the 2006 Stern Review and the release of the IPCC Assessment Report Four in 2007. Important parallel policy initiatives include the 2005 Millennium Ecosystem Assessment, the UN Millennium Development Goals (although they do not explicitly mention climate change) and the 2007 UNDP Human Development Report, which deals exclusively with the impacts of climate change and adaptation options for developing countries. The shift in mood has changed the debate about how we survive the impact of climate change as well as how we tackle emissions of greenhouse gases.

There are many reasons why people will experience climate change differently. Those currently living in stressful situations, whether created by economic hardship, conflict or environmental change, will feel the effects first and most forcefully. Emissions of greenhouse gases need to be reduced, ways of extracting greenhouse gases from the atmosphere need to be found and human capacity to cope with the impact of climate change must be improved. These are all enormous challenges, which cannot be put off. Without simultaneous action on all three fronts, there will be increased suffering both for those experiencing climate change now and those who will experience it in the future.

Putting our collective heads in the sand is not an option. It's time to take some very large and potentially world-changing decisions. When should we take action on climate change? Who should pay for adaptation to climate change? Who should take action on climate change?

Delaying taking action on global climate change will delay when we need to start paying for solutions to the problem. However, delay will put us in a position of greater information about climate change impacts – information about the exact nature of the risks and the specific areas that will suffer most. Taking anticipatory action could mean spending precious, and limited, resources in dealing with impacts that do not worsen over time or on adaptations that offer little more than short-term palliative care. However, delaying action could mean that many who are already suffering the impacts of climate change will not be helped and many, many more will be exposed to worsening impacts. And it could mean accepting significant losses (of livelihoods, cultures, homes and even lives) in the short term. These are not decisions just for politicians; they are decisions we all need to think about and take part in through the public debates, focus groups and other forms of participatory democracy on which politicians in many countries increasingly rely to form policy.

Who should pay for adaptation action? Should we leave those who are experiencing the impacts to fend for themselves ('it's just their bad luck')? Or is there a moral and ethical imperative for everyone to help each other? Should resources be redistributed from those who are not affected (or who may even be benefitting from climate change), to those who are (or will be) suffering and barely able to cope?

Any development project lasting more than five to ten years will be affected by climate change. In developing countries, significant proportions of national income are invested in large-scale infrastructure projects such as water supply, road networks, ports and airports. In November 2008, South Africa's state-owned power utility (Eskom) took out a US$500 million loan with the African Development Bank to help fund a seventeen-year expansion programme designed to double the country's electricity-generating capacity. If this, and other large projects, are not 'climate-proofed' – modified to withstand the impacts of climate change – the benefits to South Africa (and other countries) will be significantly diminished as the projects are likely to have a much shorter life.

And who should be taking action? Should everyone, in every country, expect their governments to look after them and protect them from the impacts of climate change or should we take responsibility and action ourselves? If we rely on our governments, their resources will be spread very thinly. Is it the responsibility of the government to act for the nation, when many individuals and businesses are very capable of looking after themselves? Is this the right way to spend government money; should it be spent on health care, education, social security, defence and other national priorities? There is no right answer to any of these questions, yet any answer will affect all of us for many years. The decisions taken will decide whether we can continue to live the way we have been living.

Reducing greenhouse gas emissions – today – is critical. Planning to reduce future greenhouse gas emissions is critical. Internationally, agreeing at what level atmospheric greenhouse gas concentrations should be to avoid dangerous climate change is critical. Yet we may not achieve any of them quickly enough. We do not know how to 'decarbonise' developed nations rapidly. Is it possible to go back to the carbon dioxide levels of the 1990s without losing developmental opportunities or investments in human development such as medical health care, universal education, welfare systems and peace-building?

Globally, we have four options:

Do nothing: given that the global carbon market is up and running and businesses are already pushing governments to regulate new energy options such as energy production from wind and solar power, we have already moved beyond this scenario.

Do a little: start adapting to climate change as needed, lobby major governments to assist developing countries and reduce emissions slowly, possibly stabilising at 650–750 ppm.

Do a lot: take rapid action to reduce emissions through major investments in renewable energy, encourage demand reductions and aim for a 2°C warming, stabilising at 450 ppm (but acknowledge that this is unlikely to be achieved, and accept 550–650 ppm and more than a warming of 4°C). At the same time invest heavily in adaptation to dangerous impacts.

Do a lot, plus avoid a crisis: take rapid action to reduce emissions through major investments in renewable energy, encourage demand reductions, introduce personal carbon emission allowances, legislate for greater restrictions on high carbon-intense industry, and aim for a stabilisation target of 500 ppm with sensible interim targets and clear developmental pathways. At the same time invest heavily in adaptation to dangerous impacts.

Perhaps an iterative strategy for mitigation is required, in which new information can be built into policy and policy-makers can re-evaluate policies from time to time. Some activists argue that we should revert to a 1950s mentality, accepting that emissions must be rationed and focusing on frugality, both of which were acceptable post-World War II. Of course, the values of another age are not always acceptable to the present; we can neither go back in time, nor determine what people will value in the future. To avoid forcing people back to historic values, they must be given space to reflect on mitigation and emissions targets and to find their own ways to reduce emissions. The best we can hope for is to build the capacity and flexibility for our societies to be able to change in a way that takes into account the challenges confronting each new generation.

Climate change impacts will be felt universally but distributed unfairly: there will be winners and losers. Adaptation to climate change must be supported and supported now. Yet in promoting urgent action, we must take care not to terrify people into inappropriate action, such as creating a 'fortress world' in which environmental refugees are trapped. Nor must we falsely amplify the risks people face and worry them into anxiety-filled paralysis. Hope is at the heart of adaptation: it is possible to cope with the impacts that will be experienced and people do have the capacity to adapt.

Businesses, large and small, and governments have a part to play in responsibly adapting to and mitigating climate change, supporting and rewarding actions by individuals and businesses. The relative roles of business and government will depend on the nature of the political and economic structure of each country. States with strong, trusted, central governments are likely to be more able to impose adaptation plans than countries with weak or less-trusted governments. Richer nations, governments and charities can provide financial and technical assistance to populations vulnerable to floods, droughts and other impacts. But should such funds be tied to national overseas development aid and made mandatory? In poorer nations, governments, NGOs and businesses can rally public support and provide a spur to innovative adaptions to climate change. Everywhere, the media have the duty to report accurately the facts of climate change and garner public support for government policy. Consumers can spend wisely to reduce greenhouse gas emissions, buy more locally-grown produce, use low-energy light bulbs, take public transport, recycle waste and think carefully about how much energy they are responsible for consuming, as well as becoming better informed about how to prepare for the impacts of climate change. Private companies can deal with their contributions to climate change and think about how they could adapt in sustainable ways. There are so many – almost endless – options. However, there are also many barriers to the adoption of these options.

The major emitting countries need to be included in any plan to manage climate change; both historically large emitters and emerging large emitters. Resources will be needed to support economic growth and development in poorer countries that want to develop with low-carbon technologies but cannot afford the research and development costs. Some resources are already available through the UNFCCC; however, these are unlikely to be sufficient.

In the short term, the world needs a successful global agreement at the UN Conference of the Parties (in Mexico, 2010). We, the public, must engage in the debates, show politicians our willingness to take action on climate change, question the assumptions others make about us and our behaviour, and inform ourselves about the international discussions. Any new international agreement on climate change will have to incorporate sensible global targets for the mid-twenty-first century and contain measures for global financing of adaptation. The impacts of climate change must be coupled with the long- and short-term shocks that 'peak oil' (the point in time when the maximum possible rate of crude oil extraction is reached), food security and energy security pose globally.

The 2008-2009 global financial crisis can perhaps teach us the value of rethinking how our political and economic systems function, who should be accountable and how global processes affect us as individuals, families and communities. The financial crisis also showed us clearly that sometimes we have to spend an enormous amount of money on something we don't want (such as the US$700 billion government bail-out to the financial sector) to avoid an even worse outcome. This might also be the case with climate change – we just haven't reached the crash-point yet.

Postscript

"Climategate": the story of email hacking at UEA

The nature of the incident that sparked the controversy

On Monday 23rd November 2009, the University of East Anglia (UEA) made a statement that the computer systems at the Climatic Research Unit had been hacked into and that personal information had been stolen and published illegally on a number of websites[1]. The Climatic Research Unit, established in 1972, had around thirty research staff and had developed a number of datasets widely used in climate research. Until that point it had been widely recognised as one of the world's leading institutions on natural and anthropogenic climate change.

Nearly five months later, a public inquiry into the affair by the House of Lords Science and Technology Select Committee, and a separate independent review headed by Lord Oxburgh, both completely exonerated all the scientists involved, clearly stating that there had been no breach of ethics, or of scientific standards. However in the intervening five months, the reputations of many climate scientists, and the reputation of climate science more generally, had been significant damaged. The emails caused widespread controversy, leading to calls for the

[1] For those who are interested in trawling through the thousands of emails, these can be found at http://www.eastangliaemails.com/index.php.

resignation of the very well-respected scientific director of the Climatic Research Unit (CRU), Professor Phil Jones. Worse still, scientists at CRU received death threats: these and the hacking incident are still being investigated by the Norfolk police.

The media's role in the affair

It could be argued that the story of the hacked emails at UEA was an entirely contrived media plot. As Dan Gardner explains in his 2009 book *Risk*, journalism schools teach about the qualities that make a story newsworthy, which usually includes: novelty, conflict, impact and human interest, and if the story fits into an existing narrative that the media is selling then all the better. The UEA hacked email story ticked all the boxes. Here was a very topical story about climate change – already a well developed media narrative – in the build up to the well-publicized Copenhagen meeting in 2009. It could be spun into a story about manipulative scientists lying about information that had potentially huge impacts on people around the world. It could be presented as in-fighting between scientists, public lies, duplicity, and deception. In short, it made the perfect media story. The media scrambled to be the first to claim that this was evidence that climate change was not happening. Headlines quickly appeared such as: "Hacked email Is New Fodder for Climate Dispute" (New York Times, Friday 20th November 2009).

The rise of the anti-climate change lobby and their use of the e-mails to attack climate science in general

Despite the complete exoneration of all involved, the media circus around the hacked emails continues to be used by the anti-climate change lobby as evidence of the myth of climate change. The media outputs have themselves become a tangible justification for climate denial. Climate deniers now argue that both inquiry reports are simply a cover-up. For example, in April 2010 after both inquiries on the hacked emails had released their reports, a *Daily Telegraph* journalist clearly expressed his opinion that the inquiries appear to '... share UEA's principled belief that the best response to the skullduggery exposed in the Climategate emails is to whitewash, whitewash, whitewash ...'. Research recently published in the journal *Political Behavior* in 2010 by Brendan Nyhan and Jason Reifler suggests that when the media publishes corrections to previously inaccurate stories (i.e. climate scientists didn't hide data) this correction often serves to reinforce the ideas of those with extreme views who believed the initial conspiracy story or slur to be true. Thus, despite the two inquiries showing that the accusations were false, climate deniers refuse to believe the inquiry findings, and appear vindicated that climate change is a myth!

Responses of various organisations, including inquiry reports

With hindsight, the responses by various organisations to the hacked emails were highly detrimental to climate science. Shortly after the hack into UEA, climate scientists seemed to reiterate tirelessly that climate change was really happening, without explaining the substance of the emails. No climate scientists stood up to clearly explain that 'Mike's nature trick' was a shorthand term for an accepted scientific method for managing datasets from different sources that did not agree. No climate scientist opened the debate on the issue of peer review of academic journals, and how to address the variable quality of papers published in different scientific journals. No scientists stood up and talked about the pressures on academics to publish so they are likely to clean their data adequately to analyze themselves, but unlikely to invest the time and resources to clean the data adequately for others to use. No climate scientist asked a media reporter 'have you never sent a disparaging email in confidence that afterwards, with hindsight, you wish you had never sent?' to remind the interviewer that many of us have said things in private emails that we would not wish to see published. The deafening silence from the academic community provided an own goal to climate deniers. Combined with the institutional apathy and poor public relations skills of the University of East Anglia, Professor Phil Jones was essentially hung out to dry.

Surprisingly, environmentalists also went gunning for Phil Jones. In a series of articles in the Guardian, George Monbiot called for Phil Jones, public relations people at UEA, and various other academics to stand down, assuming that the edited emails that became publicly available were evidence of scandal. Overall the media and institutional handling of the hacked emails was highly damaging for all the wrong reasons. It led to confusion among the public as to whether climate change was a real phenomenon. It leached political will to address climate change, after years of investing resources to raise the profile of the issue and to garner support for action. It was also bad for public trust in scientists. One wonders if the resulting fall-out from the hacked emails would not have occurred had the affair been handled differently by the media, UEA, and climate scientists generally.

What the affair means for climate research and science in general

Now that the event has passed and other more topical narratives have been spun by the media, it is time to reflect on some of the deeper issues that the email hacking incident raises. One is the peer-review process which scientists use to control the quality of academic publications. All scientific findings that are submitted for publication in a scientific journal must first be reviewed by at least two, and up to four, peers (i.e. scientists in other institutions who judge whether the research and findings are of sufficiently high quality to be published). Peer reviewers have three main options:

i) to reject the submitted paper due to inadequate quality
ii) to return the paper to the author recommending publication after improvements to presentation or analysis are made to bring the paper up to publishable quality
iii) to accept the paper as is, or with minor typographical amendments.

This process is known as the *peer review process*. In an era when scientists are expected to publish a higher proportion of their work, more journals are emerging to meet this demand. As a result, there are more papers to review, which means that all academics find a higher proportion of their time is taken up undertaking peer reviews. This can mean that sometimes inadequate time is given to the review process, and that papers slip through this voluntary review system. On the occasions when poor quality papers are published, other academics will critique the work and it will be discredited over time. However, for those not familiar with the peer-review process, it can appear that scientists keep changing their minds or there is dissent over an uncontroversial issue.

We are also left with the realisation that there is no such thing as a personal email. Since the email accounts at UEA were hacked, there have been at least three attacks against other UK universities, although adequate security systems appeared to prevent data extraction. Anyone who sends an email can no longer assume that this email will not be published on the internet. Before hitting 'send' we should all think about whether we would like the rest of the world see and comment on the email we have sent.

The final outcome from the UEA email saga has been the recognition of the importance of scientific candidness. Specifically, what is needed to ensure that all outputs from publicly funded science are accessible? Here again, there are issues worthy of more thought. How should we balance the intellectual property rights of the researcher designing the study and gathering the data against the rights of other scientists to use their data? How much money should we divert from research and education into support to transform all data into publicly accessible information? To what degree should members of the public have the right to download, review, and critique data upon which scientific arguments are based – without the skills or knowledge to understand what the data are saying? We do not have the answer to these questions, however we will look forward to a public debate to discuss these issues – should the media feel they are worthy of discussion.

Postscript

Copenhagen 2009 and beyond

Introduction

The United Nations Climate Change Conference was held in Copenhagen between 7th and 19th December 2009. Around 115 world leaders attended the meeting, alongside more than 40,000 people from government, non-governmental organisations, intergovernmental organisations, faith-based organisations, the media, and UN-accredited organisations (ENB, 2009). For many, the meeting was considered a window of opportunity and 'the most important meeting of our lives'. Others warned that expectations were too high and that, at most, a loose political agreement would be achieved that would set the stage for a legally-binding agreement to be reached in Mexico in 2010. The over-arching objective of the Copenhagen meeting was to reach agreement on levels of greenhouse gas emissions reductions by 2020, and to set targets for 2050.

While, in many ways, the meeting was historic, its outcomes were a poor reflection of our collective ability to address global climate change. This chapter reflects on the pre-Copenhagen process, the outcomes, and what may lie ahead beyond Copenhagen.

Hopes for Copenhagen

Since the launch of the United Nations Bali Roadmap on climate change in December 2007, the hopes of many policy-makers, NGOs, civil society and private companies were pinned on achieving meaningful and long-term global action on climate change at Copenhagen 2009. A two-year negotiating process between the Bali and Copenhagen meetings raised hopes that the 2009 conference would be a chance for the global community finally to deliver commitments for a strong agreement on climate change. However, the process also illustrated major divisions between richer and poorer countries.

Earlier in 2009, discussions had stalled on the issue of substantial cuts in emissions, when poorer countries had urged richer countries to commit to ambitious targets. However, the response was that without the United States and major developing countries such as India and China, an agreement on climate change would not be meaningful (ENB, 2009).

Specific key policy outcomes were hoped for, including agreement on:

- Ambitious medium-term cuts in carbon dioxide emissions by developed countries
- Clarity on mitigation actions by major developing countries
- Substantial financing and action to improve adaptation
- Short and long-term finance
- Technology transfer
- Reducing emissions from deforestation and degradation in developing countries
- A shared vision on how to achieve long-term global cooperative action, by 2050

In the run-up to Copenhagen, many fringe voices expressed varied views on priorities for climate change. Among civil society, there were mounting fears about potentially dangerous climate change, illustrated by campaigns such as 'TckTckTck' – a coalition of 15 million people (both individuals and civil organisations) fighting climate change. The fears focussed on the idea of a 'tipping point' in the Earth's climate system; a theoretical point that occurs when feedback that has not been accounted for in climate models, such as the ocean system, leads to unexpected cumulative climate change effects. To avoid this, some researchers suggest we need to keep global warming below 2°C. This requires a reduction in greenhouse gases to 350 parts per million, translating to cuts in emissions of 40% below 1990 levels by 2020. To achieve these cuts, richer developing countries would also need to make commitments to cut greenhouse gases to below current emissions. However, this is a highly controversial point.

From businesses, the Copenhagen meeting was expected to deliver a flexible framework for future negotiations and medium-term emissions targets, and deploy mass investments in research and development of technologies such as carbon capture and storage. Improvements to the Clean Development Mechanism and financial aid to poor countries in the realm of $200 billion a year were called for.

Forests also play an important role in the climate change debate, because greenhouse gas emissions from deforestation contribute up to 20% of global greenhouse gas emissions. Many had high hopes that the negotiations would result in agreement on the REDD mechanism (Reduced Emissions from Deforestation and Degradation) which is designed to benefit national governments through extending the ecological services of forests to the global carbon market and assisting poorer communities to mitigate and adapt to climate change. For example, governments would be paid to protect forests in exchange for carbon credits and could then distribute these funds for development.

Before the Copenhagen meeting, some warned that high expectations would ultimately lead to the collapse of the process. Jeffrey Sachs, one of the present day's most well-recognised development economists, indicated in an interview with EarthSky (http://earthsky.org/human-world) that countries would fail to reach a comprehensive agreement. He argued that diplomats with limited knowledge of technological solutions conduct political negotiations wrongly. In other words, there is a gap between political process and practical technical solutions. Sachs, and many others, predicted that at most a political agreement would be achieved in Copenhagen that would pave the way for a future legally-binding agreement.

The Copenhagen process and outcomes

What happened at Copenhagen came close to Sachs's predictions. The meeting was a disappointment to some government negotiators and many civil society groups. The negotiations were mired in controversy over transparency and bogged down by time-consuming procedural issues. In particular, there were disagreements over the way that the Danish Presidency, which was hosting the event, handled the negotiations. High-level negotiations between presidents and prime ministers resulted in a political agreement, the 'Copenhagen Accord'. The announcement of the Accord to the international press by the President of the USA, Barack Obama, was criticised as premature and 'out of sync' with the political procedures of the UN.

The closing plenary of the conference saw some acrimonious disputes and a small number of delegates were especially outraged. In particular, some developing countries felt that the Accord was tantamount to suicide for vulnerable countries (ENB, 2009). The UN Secretary-General, Ban Ki-Moon, had to facilitate fraught negotiations to ensure that countries could reach an agreement on the Accord. In the end, countries took note of the Copenhagen Accord and agreed, on a voluntary basis, to register their support for it and submit targets by 31 January 2010 (ENB, 2009). Reportedly, since Copenhagen, 'more than 100 countries have associated themselves with the Accord and as a result of the targets and actions put forward, around 80% of emissions are covered by the agreement' (DECC, 2010:7).

The Accord included elements of a framework:

- An agreement on a 2°C target (with a reference to limiting temperature increase to below 1.5°C)
- Financing for climate change and development with pledges of US$ 30 billion to the developing countries between 2010-2012, rising to US$ 100 billion per year by 2020, to help poor countries tackle climate change
- Financial pledges and a mechanism on forests (Reddplus)
- Establishment of new bodies: a High-Level panel under the UN COP to study financial provisions and implementation, the Copenhagen Green Climate Fund, and a Technology Mechanism
- Nations to submit their emissions pledges on a voluntary basis to the UN. No legal commitments were reached on reducing emissions and no quantification of a long-term global emissions reduction goal or specific timings for peak emissions
- Mitigation actions by developing countries include no mention of quantified emissions reductions, but there is reference to monitoring and verification procedures in which developing countries report to the UN through their national communications procedures. Text on provisions for international consultations on monitoring and verifying national mitigation actions has yet to be fully explained (IISD, 2009)
- A review of the operation of the Accord by 2015

Beyond Copenhagen

Did Copenhagen succeed or fail?

The Accord is a record of how far the negotiations were able to achieve their objectives. First, no substantial commitments to medium-term cuts were achieved, yet willing countries reached agreement on voluntary targets. Since the Copenhagen meeting, it seems that over 100 countries have associated themselves with the Accord. Second, developing countries agreed to report on some mitigation actions, again voluntarily. Third, finances were pledged, not at the scale of $200 billion per annum, but to half that level by 2020; pledges were given for short-term finance. Finally, in addition to the Accord, an important agreement that came out of Copenhagen was to continue to support the work of the two technical bodies working on long-term co-operative action and on further commitments under the Kyoto Protocol for one more year.

The Accord is deemed better than it seemed in the immediate aftermath of Copenhagen. However, critics of global climate policy are not so optimistic. The *Hartwell Paper*, co-written by fourteen international scholars suggests that global climate policy crashed in 2009, resulting in the need seriously to reconsider alternative avenues for climate policy. The paper states that the Kyoto Protocol approach failed to achieve real world global greenhouse gas emissions reductions, due to structural flaws in the UN system and failures to systematically understand the nature of climate change as a policy issue (Prins et al., 2010).

The next UN climate meeting will take place in Cancun, Mexico in 2010. This meeting has, so far, received much less attention than Copenhagen. However, in analysing the present situation there are other factors to consider, such as the financial crisis in Europe and the threats to the Euro, and President Obama's problems with climate and energy policy in the US Senate, particularly in view of the congressional elections that will take place in the autumn of 2010. These have taken centre stage in the world's media. In the mean time, social movements continue to pressurise governments to take action. For example, the 'World's People's Conference on Climate Change and the Rights of Mother Earth' held in Bolivia in April 2010 showed how social movements continue to advocate climate action outside the formal political climate negotiation processes. Also, governments continue to work on strategies for a multilateral agreement. For instance, the UK has produced a strategy on climate change action: *Beyond Copenhagen: The UK Government's International Climate Action Plan* (DECC, 2010). This strategy illustrates how governments continue to advocate their commitment to movement towards low-carbon and resilient economies.

In conclusion, the financial crisis, which began in 2008, has brought one other important lesson home: while international approaches, whether to climate change or financial markets, are vital in today's globalised world, there is also a need to establish strategies and mechanisms at the domestic level and in civil society to buffer against uncertain events or stalled global political processes. A multi-layered approach is necessary to ensure that our hopes do not hinge, precariously, on the success or failure of one set of processes.

Emily Boyd and Emma L. Tompkins
April 2010

Further reading

General reading on climate change

Tim Flannery's (2006) *The Weather Makers: How Man Is Changing the Climate and What It Means* focuses on the idea that the world is at a global climatic tipping point. Flannery emphasises the accumulation of cataclysmic events, such as the most powerful El Niño ever recorded, the most devastating hurricane in two hundred years, the hottest European summer on record, one of the worst storm seasons ever experienced in Florida, the extinction of one in five species on the planet and more. Flannery's account is an urgent warning of what societies must do to prevent a disastrous future.

Mike Hulmes's (2009) *Why We Disagree About Climate Change: Understanding Controversy, Inaction and Opportunity* focuses on the environmental, cultural and political dimensions of climate change, which are currently reshaping the way that societies see themselves and the planet. Hulme provides an account of the emergence of climate change as a social phenomenon from different theoretical perspectives: science, economics, faith, psychology, communication, sociology, politics and development to explain why we disagree about climate change.

Mark Lynas's (2007) *Six Degrees: Our Future on a Hotter Planet* ventures along similar lines, extrapolating from what climate scientists have to say about the impact of climate change over the next hundred years and outlining what to expect from a warming world, degree by degree. He provides a horrifying preview of what would happen at 6°C of warming – the extinction of mankind – inferred from palaeo-archaeological records.

James Martin's (2007) *The Meaning of the 21st Century – A Vital Blueprint for Ensuring our Future* focuses on grand schemes of revolutionary change and high technology solutions. He notes that the history of revolution is pockmarked with violence, winners and losers. Change will occur on many scales; generating Utopia for some but dystopia for most.

Nicholas Stern's (2009) *A Blueprint for a Safer Planet: How to Manage Climate Change and Create a New Era of Progress and Prosperity* focuses on the economic management of investment and growth from the perspective of both adaptation and mitigation. Stern examines the history of the problem, the dangers and the costs of emissions reductions; as well as discussing the challenges of adaptation and ethics, the policies needed to reduce emissions at different societal levels, and providing a structure for a new global climate deal.

Gabrielle Walker and Sir David King's (2008) *The Hot Topic: How to Tackle Global Warming and Still Keep the Lights On* provides pragmatic and useful insights on how to tackle climate change. In addition to their mature perspective on climate change, much of the book focuses on climate science and mitigation solutions.

Chapter 1

IPCC (2007) *Climate Change 2007: Impacts, Adaptation and Vulnerability*. Contributions of Working Group II to the Fourth Assessment Report of the Intergovernmental Panel on Climate Change, M.L. Parry, O.F. Canziani, J.P. Palutikof, P.J. van der Linden and C.E. Hanson, eds., Cambridge University Press, Cambridge, UK, p. 976

Liverman, D. (2007) 'Assessing impacts, adaptation and vulnerability: Reflections on the Working Group II Report of the Intergovernmental Panel on Climate Change.' *Global Environmental Change*, 18(1): 4–7

Liverman, D. (2007) 'From Uncertain to Unequivocal. The IPCC Working Group I Report: Climate Change 2007 – The Physical Science Basis'. *Environment*, 49(8): 36–9

Websites

Resilience (www.ukresilience.info) (accessed 14 March, 2008)

Cape Farewell (www.capefarewell.com) (accessed 14 March, 2008)

Chapter 2

Allen, M. (1999) 'Do-it-yourself climate prediction'. *Nature*, 401: 602

Bryden, Harry L., Longworth, Hannah R. and Cunningham, Stuart A. (2005) 'Slowing of the Atlantic meridional overturning circulation at 25° N'. *Nature*, 438: 655–657

Luthcke, S. B., Zwally, H. J., Abdalati, W., Rowlands, D. D., Ray, R. D., Nerem, R. S., Lemoine, F. G., McCarthy, J. J., Chinn, D. S. (2006) 'Recent Greenland Ice Mass Loss by Drainage System from Satellite Gravity Observations'. *Science*, 314: (5803) 1286–1289

Miller, C.A. and Edwards, P.N. (eds.) (2001) *Changing the Atmosphere: Expert Knowledge and Environmental Governance*. MIT Press, Cambridge, Massachusetts

Soon, W. et al. (2001) 'Modeling climatic effects of anthropogenic carbon dioxide emissions: unknowns and uncertainties'. *Climate Research*, 18: 259–275.

Stainforth, D.A. et al. (2005) 'Uncertainty in predictions of the climate response to rising levels of greenhouse gases'. *Nature*, 433: 403–406

Websites

ClimatePrediction.net (www.climateprediction.net) (accessed 14 March, 2008)

Wikipedia (http://en.wikipedia.org/wiki/Climate_model) (accessed 14 March, 2008)

Chapter 3

Blaikie, P., T. Cannon, I. Davis and B. Wisner (1994) 'At Risk: Natural Hazards, People's Vulnerability and Disasters'. Routledge, London

Christensen, J.H., Hewitson, B., A, Busuioc, A. Chen, X. Gao, I. Held, R. Jones, R.Kolli, W-T. Kwon, R. Laprise, V. Magaña Rueda, L. Mearns, C. Guillermo Menéndez, J. Räisänen, A. Rinke, A. Sarr and P. Whetton (2007) 'Chapter 11: Regional climate projections' in *IPCC Working Group I* (2007), 847–940

Confalonieri, U., B. Menne, R. Akhtar, K.L. Ebi, M. Hauengue, R.S. Kovats, B. Revich and A. Woodward (2007) 'Human health. Climate Change 2007: Impacts, Adaptation and Vulnerability'. *Contribution of Working Group II to the Fourth Assessment Report of the Intergovernmental Panel on Climate Change.* M.L. Parry, O.F. Canziani, J.P. Palutikof, P.J. van der Linden and C.E. Hanson, (eds.), Cambridge University Press, Cambridge, UK, 391–431

Easterling, W.E., P.K. Aggarwal, P. Batima, K.M. Brander, L. Erda, S.M. Howden, A. Kirilenko, J. Morton, J.-F. Soussana, J. Schmidhuber and F.N. Tubiello (2007) 'Food, fibre and forest products. Climate Change 2007: Impacts, Adaptation and Vulnerability.' *Contribution of Working Group II to the Fourth Assessment Report of the Intergovernmental Panel on Climate Change.* M.L. Parry, O.F. Canziani, J.P. Palutikof, P.J. van der Linden and C.E. Hanson (eds.), Cambridge University Press, Cambridge, UK, 273–313

Fischlin, A., G.F. Midgley, J.T. Price, R. Leemans, B. Gopal, C. Turley, M.D.A. Rounsevell, O.P. Dube, J. Tarazona, A.A. Velichko (2007) 'Ecosystems, their properties, goods and services. Climate Change 2007: Impacts, Adaptation and Vulnerability.' *Contribution of Working Group II to the Fourth Assessment Report of the Intergovernmental Panel on Climate Change.* M.L. Parry, O.F. Canziani, J.P. Palutikof, P.J. van der Linden and C.E. Hanson (eds.), Cambridge University Press, Cambridge, 211–272

Millennium Ecosystem Assessment (2005) *Ecosystems and Human Wellbeing.* Island Press, Washington D.C.

UNEP (1990) *Food security and agriculture: The impacts of climate change on agriculture.* United Nations Environment Programme Information Unit for Climate Change Fact Sheet 101. Nairobi, Kenya

Chapter 4

Anderson, K., Shackley, S., Mander, S. and Bows, A. (2005) *Decarbonising the UK. Energy for a Climate Conscious Future,* Tyndall Centre for Climate Change Research, Norwich, UK

Burton, I. and M. van Aalst (2004) *Look Before You Leap: A Risk Management Approach for Incorporating Climate Change Adaptation into World Bank Operations.* The World Bank, Washington, DC

Harremoës, P., Gee, D., MacGarvin, M., Stirling, A., Keys, J., Wynne, B. and Vaz, S.G. (eds.) (2001) *Late lessons from early warnings: the precautionary principle 1896–2000.* European Environment Agency, Copenhagen

IPCC (2007) *Climate Change 2007: Impacts, Adaptation and Vulnerability. Working Group II Contribution to the Intergovernmental Panel on Climate Change Fourth Assessment Report,* IPCC, Geneva, Switzerland

IPCC, Bert Metz, Ogunlade Davidson, Heleen de Coninck, Manuela Loos and Leo Meyer (eds.) (2005) *Carbon Dioxide Capture and Storage.* Cambridge University Press, Cambridge, UK

Mather, T. (2005) *Postnote: Carbon Capture and Storage* (CCS). March 2005 Number 238, The Parliamentary Office of Science and Technology, HM Stationery Office, London

McCarthy, J.J., Canziani, O.F., Leary, N.A., Dokken, D.J. and White, K.S. (eds.) (2001) *Climate Change 2001: Impacts, Adaptation, Vulnerability. Contribution of Working Group II.* Third Assessment Report of the Intergovernmental Panel on Climate Change. Cambridge University Press, Cambridge, UK and New York, USA

Metz, B., Davidson, O., Swart, R. and Pan, J. (eds.) (2001) *Climate Change 2001: Mitigation. Contribution of Working Group III.* Third Assessment Report of the Intergovernmental Panel on Climate Change. Cambridge University Press, Cambridge

New Economics Foundation (2004) *Up in Smoke*. New Economics Foundation, London

Tompkins, E.L., Nicholson-Cole, S.A., Hurlston, L.-A., Boyd, E., Hodge, G.B., Clarke, J., Gray, G., Trotz, N. and Varlack, L. (2005) *Surviving climate change in small islands: a guidebook*. Tyndall Centre for Climate Change Research, University of East Anglia, Norwich, UK

Chapter 5

Andresen, S. and Agrawala, S. (2002) 'Leaders, pushers and laggards in the making of the climate regime'. *Global Environmental Change*, 12: 41–51

Boyd, E., Hultman, N., Roberts, T. et al. (2009) 'Reforming the CDM for sustainable development: lessons learned and policy futures'. *Environment Science and Policy* 12(7): 820–31

Boykoff, M and Roberts, T. (2007) 'Media Coverage of Climate Change: Current Trends, Strengths and Weaknesses'. *Human Development Report* 2007/2008

Capoor, K. and Ambrosi, P. (2009) State and Trends of the World's Carbon Markets 2009. Accessed July 2009 at http://wbcarbonfinance.org/docs

ClimateFundsUpdate (2009) *Current Climate Funds List*. London: Overseas Development Institute and Heinrich Böll Foundation. Accessed September 2009 at www.climatefundsupdate.org/graphs-statistics/pledged-deposited-disbursed

Directgov. Climate change: a quick guide and Carbon Calculator at www.direct.gov.uk accessed 24 June 2008

Grubb, M. and Vrolijk, C. (eds.) (1999) *The Kyoto Protocol: a guide and assessment*. The Royal Institute of International Affairs/Earthscan, London

Gupta, J. (2000) *On Behalf of My Delegation. A Survival Guide for Developing Country Climate Negotiations*. Center for Sustainable Development of the Americas, Washington DC

Martin, J. (2006) *The Meaning of the 21st Century A Vital Blueprint for Ensuring our Future.* Transworld Publishers, London

Moser, S.C. and Dilling, L. (eds.) (2007) *Creating a Climate for Change: Communicating Climate Change and Facilitating Social Change.* Cambridge University Press, Cambridge, UK

OECD (2009) 'Development Aid at Its Highest Ever Level in 2008'. *Organisation for Economic Co-operation and Development.* Accessed June 2009 at www.oecd.org

Rayner, S. and Malone, E.L. (1998) *Human Choice and Climate Change. Volume 4: What Have We Learned?*, Battelle Press, Columbus, OH

Revkin, A. (2007a) 'Poorest Nations Will Bear Brunt as World Warms'. *New York Times.* New York, NY, 1 April 2007

Revkin, A. (2007b) 'The Climate Divide: Wealth and Poverty, Drought and Flood: Reports from Four Fronts in the War on Warming'. *New York Times.* New York, NY, 3 April 2007

Schneider, S.H., Rosencranz, A. and O'Niles, J.O. (eds.) (2002) *Climate Change Policy a Survey.* Island Press, Washington DC

UNFCCC 2009 CDM Statistics. Accessed 23 June 2009 at www.cdm.unfccc.int/statistics/index.htm

Chapter 6

Dessai, S., Adger, W.N., Hulme, M., Turnpenny, J., Köhler, J. and R., W. (2004) 'Defining and experiencing dangerous climate change'. *Climatic Change*, 64: 11–25

Dow, K. and Downing, T.E. (2007) *The Atlas of Climate Change: Mapping the World's Greatest Challenge.* Earthscan, UK

Giddens, A. (2009) *The Politics of Climate Change.* Polity Press, Cambridge, UK

Grothmann, T. and Patt, A. (2005) 'Adaptive capacity and human cognition: the process of individual adaptation to climate change'. *Global Environmental Change*, 15 (3): 199–213

Hulme, M. (2009) *Why We Disagree about Climate Change:*

Understanding Controversy, Inaction and Opportunity. Cambridge University Press, Cambridge, UK

Mastrandrea, M. D. and Schneider, S. H. (2004) 'Probabilistic integrated assessment of dangerous climate change'. *Science*, 304, 571–5

O'Brien, K. L. and Leichenko, R. M. (2003) 'Winners and Losers in the Context of Global Change'. *Annals of the Association of American Geographers*, 93, 89–103

Pidgeon, N., Kasperson, R. E. and Slovic, P. (2003) *The Social Amplification of Risk*. Cambridge University Press, Cambridge, UK

Potts, J. S. (1999) 'The non-statutory approach to coastal defence in England and Wales: Coastal Defence Groups and Shoreline Management Plans', *Marine Policy*, 23, 479–500

Schipper, E.L. and Burton, I. (eds.) (2009) *The Earthscan Reader on Adaptation to Climate Change*. Earthscan Reader Series. Earthscan, London

Stern, N. (2009) *A blueprint for a safer planet: how to manage climate change and create a new era of progress and prosperity*. Bodley Head, London

Stern, N. (2006) *The Economics of Climate Change: The Stern Review*, London, Cambridge University Press, Cambridge, UK

Tompkins, E.L., R. Few and K. Brown (2008) 'Scenano-based stakeholder engagement: Incorporating stakeholders' preferences into coastal planning for climate change'. *Journal of Environmental Management* 88(4): 1580–92.

UNEP (2002) *Global Environment Outlook 3. Past, present and future perspectives*. Earthscan, London

Postscript

DECC (Department of Energy and Climate Change). 2010. *Beyond Copenhagen: The UK Government's International Climate Change Action Plan*. The Office of Public Sector Information. Her Majesty's Stationary Office (HMSO), United Kingdom.

ENB (Earth Negotiations Bulletin). 2009. *Summary of the Copenhagen Climate Change Conference, 7–19 December 2009*, Vol.12, No.459. International Institute for Sustainable Development.

Prins G., I. Galiana, C. Green, R. Grundmann, M. Hulme, A. Korhola, F.Laird, T. Nordhaus, R. Pielke Jr., S. Rayner, D. Sarewitz, M. Shellenberger, N. Stehr, H. Tezuka. 2010. *The Hartwell Paper: A new direction for climate policy after the crash of 2009.* London School of Economics and Oxford University.

Index